T0321249

# SYNCHRONIZATION
# DESIGN FOR DIGITAL SYSTEMS

# SYNCHRONIZATION DESIGN FOR DIGITAL SYSTEMS

by

**Teresa H. Meng**
Stanford University

*with contributions by*

*David Messerschmitt*
*University of California, Berkeley*

*Steven Nowick*
*Stanford University*

*David Dill*
*Stanford University*

**Kluwer Academic Publishers**
**Boston/Dordrecht/London**

**Distributors for North America:**
Kluwer Academic Publishers
101 Philip Drive
Assinippi Park
Norwell, Massachusetts 02061 USA

**Distributors for all other countries:**
Kluwer Academic Publishers Group
Distribution Centre
Post Office Box 322
3300 AH Dordrecht, THE NETHERLANDS

**Library of Congress Cataloging-in-Publication Data**

Meng, Teresa H.
  Synchronization design for digital systems / by Teresa H. Meng.
    p.   cm. — (The Kluwer international series in engineering and
  computer science ; SECS 123)
  Includes bibliographical references and index.
  ISBN 0-7923-9128-4
  1. Timing circuits—Design and construction.   2. Digital
integrated circuits—Design and construction.   3. Discrete-time
systems.   I. Title.   II. Series.
TK7868.T5M46   1991
621.381 '5—dc20                                          90-49357
                                                              CIP

*Printed on acid-free paper.*

Printed in the United States of America

# CONTENTS

# 4  SELF-TIMED PROGRAMMABLE PROCESSORS  67

# 7 AUTOMATIC VERIFICATION 147

**Steve M. Nowick and David L. Dill**

# PREFACE

Synchronization is one of the important issues in digital system design. While other approaches have always been intriguing, up until now synchronous operation using a common clock has been the dominant design philosophy. However, we have reached the point, with advances in technology, where other options should be given serious consideration. This is because the clock periods are getting much smaller in relation to the interconnect propagation delays, even within a single chip and certainly at the board and backplane level. To a large extent, this problem can be overcome with careful clock distribution in synchronous design, and tools for computer-aided design of clock distribution. However, this places global constraints on the design, making it necessary, for example, to redesign the clock distribution each time any part of the system is changed.

In this book, some alternative approaches to synchronization in digital system design are described and developed. We owe these techniques to a long history of effort in both digital system design and in digital communications, the latter field being relevant because large propagation delays have always been a dominant consideration in design. While synchronous design is discussed and contrasted to the other techniques in Chapter 6, the dominant theme of this book is alternative approaches.

One of the problems with this field is careless and even inconsistent use of terminology. A particular difficulty is that the fields of digital systems and digital communication use inconsistent terminology. Thus, in Chapter 2 a terminology and taxonomy of synchronization approaches is defined, and is then applied consistently throughout the book. Frankly, we hope to influence workers in this field to use this unambiguous terminology in the future.

The greatest part of the book is devoted to *anisochronous* interconnect, which has traditionally been called self-timed or asynchronous interconnect in the prior literature. This development is based on Teresa Meng's Ph.D. thesis, in which she systematically developed this technique, building on the earlier work of many researchers, and in particular developed automatic design synthesis techniques. The result is a practical approach to anisochronous design that guarantees by construction correct speed-independent operation.

In Chapter 6, a class of *isochronous* design approaches is developed. This class is based on clock distribution, and includes the traditional synchronous methodology. It also includes *mesochronous* interconnect, which is an intriguing alternative to anisochronous design. It has many of the benefits of anisochronous, such as delay-independence. In fact, quantitative comparisons are developed that show an inherently lower degradation in throughput with increasing interconnect delay for isochronous methods as compared to anisochronous methods.

All the synchronization methods developed in this book have been successfully applied in practice, so they are real and worthy of consideration by system designers. For example, mesochronous interconnect is widely used in the worldwide digital telecommunications networks, a very large digital system indeed! We hope that this book will influence future digital system designers to consider synchronization methodologies other than the traditional synchronous design. We will consider our efforts to be fruitful if some of the techniques we describe here make their way into commercial system designs. This book hopefully will provide prospective designers with many of the insights and tools they need to make this a reality.

*David G. Messerschmitt*

Berkeley, California

# ACKNOWLEDGEMENTS

I would like to thank my advisor, Prof. David Messerschmitt, for always having confidence in me. His encouragement changed my attitude toward my work and his generosity and warm personality set up my idea of a true scholar. The five years at Berkeley had been the most rewarding years in my life and Dave had provided me with such a pleasant environment that I not only learned, I enjoyed it too.

My other advisor, Prof. Bob Brodersen, helped me in many different ways. It is always very inspiring to discuss technical issues with Bob. I thank him for forcing me to appreciate the problems in the real world and helping me to develop knowledge in IC design.

My friends in our group, just to name a few, Vijay Madisetti, Keshab Parhi, Wen-Long Chen, and Biswa Ghosh had made my late nights at school very pleasant ones. I'd like to give my special thanks to Edward Lee for always being encouraging and supportive. He also recalled my hidden interest in visual arts and showed me that we can be an engineer and an artist at the same time.

I would like to thank to Mr. Steve Nowick for verifying the circuits presented in this book and his valuable comments on how Boolean functions can be represented using logic gates in a speed-independent way. I would also like to thank to Dr. Gordon Jacobs for his collaboration on this work, and to all the people who were interested in my work and shared with me their intelligence and experience. I am grateful to their warm critics and friendships; to name a few: Prof. David Dill at Stanford University, Prof. Chuck Seitz, Prof. Allain Martin, and Mr. Steve Burns at California Institute of Technology, Prof. S. Y. Kung at Princeton University, and Prof. Carlo Sequin at UC Berkeley.

# SYNCHRONIZATION
# DESIGN FOR DIGITAL SYSTEMS

# 1

# INTRODUCTION

*"We might say that the clock enables us to introduce a discreteness into time, so that time for some purposes can be regarded as a succession of instants instead of a continuous flow. A digital machine must essentially deal with discrete objects, and in the case of the ACE [automatic computing engine] that is made possible by the use of a clock. All other digital computing machines except for human and other brains that I know of do the same. One can think up ways of avoiding it, but they are very awkward."*

*Alan Turing, 1947*
*Lecture to the London*
*Mathematical Society*

The issues in designing computing machines, both in hardware and in software, have always shifted in response to the evolution in technology. VLSI promises great processing power at low cost, but there are also new constraints that potentially prevent us from taking advantage of technology advances. The increase in processing power is a direct consequence of scaling the digital IC process, but as this scaling continues, it is doubtful that the benefits of faster devices can be fully exploited due to other fundamental limitations. It is already becoming clear that the system clock speeds are starting to lag behind logic speeds in recent chip designs. While gate delays are well below 1 nanosecond in advanced CMOS technology, clock rates of more than 50 Mhz are difficult to obtain and where they have been attained require extensive design and simulation effort. This problem will get worse in the future as we integrate more devices on a chip and the speed of logic increases further.

Asynchronous design, which does not require an external clocking signal, gives better performance than comparable synchronous design in situations for which global synchronization with a high speed clock becomes a limiting factor to system throughput. Automatic synthesis and the ability to decouple timing considerations from the circuit design make this approach particularly attractive in reducing design effort when the system becomes complex. The simplicity of design plus the potential performance advantages motivate our interest in designing digital systems using an asynchronous approach, and thus our attention on the synchronization design for such systems.

In this book, we will describe a systematic procedure an automated algorithm of synthesizing self-timed synchronization circuits from a structural specification. Two computer-aided design tools, the synthesis program and an event-driven simulator (designed especially for emulating hardware operations built of asynchronous components) have been developed as an effort to facilitate design automation. The design procedure has been applied to both programmable and dedicated-hardware architectures. Synchronization alternatives other than asynchronous handshakes will also be covered and a taxonomy of synchronization will be given. To complete our discussion on the synchronization design for digital systems, the theory and practice of using automatic verification as a synthesis technique will be discussed and the design of a correct arbiter will be given at the end of this book.

## 1.1. ASYNCHRONOUS AND SYNCHRONOUS DESIGN

One important design consideration that limits clock speeds is clock skew [1], which is the phase difference of a global synchronization signal at different locations in the system. When multi-phase clocks are used, a dead time or *non-overlap* time is placed between clock phases in order to absorb clock skew on a chip. Increasing the non-overlap time between clock phases prevents incorrect operation due to clock skew but reduces the time available for computation. Clock skew can be reduced with proper clock distribution [2,3], but because of this global constraint high-performance circuitry is usually confined to a small chip area. Clock skew has become a major design consideration in wafer scale integration and multi-chip board level design [4].

The asynchronous design approach eliminates the need for a global clock and circumvents the problems due to clock skew. At the chip level, layout and simulation effort is greatly reduced since there is no global timing [5]. At the board level, systems can be easily extended without problems in global synchronization [6,7]. This is particularly important for designs using pipelined architectures, where computation can be extended and sped up using pipelining without any global constraint on the overall system throughput, whereas the global clock used in synchronous design will eventually limit either the system throughput or the physical size.

Asynchronous design has not been extensively used. Among its problems, hazard and race conditions embedded in asynchronous logic and the inherent metastability phenomenon in arbiters are often mentioned. The biggest difficulty is that logic has been slow relative to easily obtainable clock speeds, in which case the overhead in asynchronous interface logic networks (commonly called *handshake circuits*) is substantial.

Significant work has been done on designing asynchronous logic using state flow graphs, or truth tables [8]. Asynchronous circuits with bounded delay elements have been investigated, often referred to as the Huffman-modeled circuits [9]. Erroneous behavior at the output of a combinational network can occur if the signal values on more than one line in the network are changed, called a *race condition*. This erroneous behavior is transient because the output can only be temporarily in error in a feedback-free combinational network. With the assumption of bounded gate delays, temporarily erroneous outputs can be eliminated by designing a hazard-free network, usually by including all the prime implicants to the network and by restricting changes of inputs to a single variable at a time [10]. In sequential networks, race conditions can lead to steady-state errors, causing malfunction and incorrect outputs. If all the gate delays are bounded, it is possible to obtain correct sequential behavior by introducing redundant states and making the delay elements in each feedback loop sufficiently long for all combinational circuits to settle down before transmitting a change of state [11,10,12]. However, the design procedure is error-prone and the restriction on input signals puts a severe limitation on the asynchronous circuits that can be realized.

Asynchronous circuits with unbounded gate delays have used data detectors and spacers [8,13] or multi-valued circuits [14]. Coding schemes [15,16,17] have been used to encode data lines so that data detectors will signal when outputs become stable, requiring at least twice the hardware compared with usual combinational logic. Spacers and coding reduce the hardware efficiency to less than 50% because half of the processing cycle is used for reseting. In this book, we will describe a different synthesis approach as compared with the classic asynchronous circuit design, in which concurrency, or performance, is our synthesis criterion.

Another difficulty often mentioned in asynchronous design is the performance uncertainty due to metastability [18]. On example is an arbiter, which grants mutually-exclusive accesses to a common resource requested by multiple users. Simultaneous requests can drive arbiters into a metastable state, where the output is undefined for a nondeterministic period of time [19,20,21,22,23,24]. The uncertainty results in either an indeterminate response time if correct outputs are required, or incorrect outputs if a maximum allowable response time is enforced.

Metastability is considered inevitable in any circuit design of this sort. Metastability is not unique to asynchronous systems, since any synchronous system that allows non-deterministic operations such as fair mutually-exclusive memory accesses incorporates metastable circuits. At the circuit level, any cross-coupled inverter pair potentially exhibits metastability, but in a synchronous design the clock rate is usually designed to be low enough that the latching signal always falls behind the data transition. The latching signal, or the clock, and the data signal can be seen as two independent input sources to the inverter coupled pair, thus introducing the uncertainty in timing. In asynchronous design, the latching signal can be derived directly from the data signal so that latching will not take place until the data signal is valid. Asynchronous design therefore does not introduce *more* undecidable timing concerns as is often suspected.

## 1.2. MOTIVATION FOR ASYNCHRONOUS DESIGN

With the advances in technology, the area penalty of using asynchronous design is becoming less significant, and the performance penalty is being similarly reduced. Several recent dedicated-hardware designs have achieved high performance through the use of self-timed circuits [25,26]. Since the asynchronous design approach represents a significant departure from current design methods, we will examine further the motivations for taking this approach.

### 1.2.1. Scaling and Technological Reasons

While scaling the IC process reduces the area requirements for a given circuit, more circuitry is usually placed on a chip to take advantage of the extra space and to further reduce system costs [27]. Therefore for a global signal such as a clock, the capacitive load tends to stay constant with scaling. Clock skew is a direct result of the *RC* time constants of the wires and the capacitive loading on clock lines. As the wire width is scaled down, the two dimensional effects of its capacitance to the substrate and surrounding layers become dominant [28]. Hence the capacitance associated with the interconnect layers cannot scale below a lower limit. As devices get smaller, the clock load represents a larger proportional load to the scaling factor and more buffering is needed to drive it, complicating the clock skew problem.

To increase the clock frequency, the rise time of the clock signal must decrease accordingly. To charge the load on the clock line in a shorter time requires higher peak currents. The higher peak currents aggravate the problem of *electromigration*, which causes aluminum (AL) wires to open circuit over time if the current density is greater than about $1 \, mA/\mu^2$ [29]. To avoid having prohibitively wide AL wires, materials such as tungsten (W) can be used to avoid the electromigration problem. Tungsten has a sheet resistance of about three to five times greater than AL [27].

Consequently, as the scaling of the IC process continues, the interconnect capacitance does not scale proportionally and the wire resistance will tend to increase from the use of other materials. Calculations show that it would take roughly $1 \, nsec$ for a very large transistor ($1/gm = 50\Omega$) to charge a capacitance of $10 \, pf$ in advanced submicron CMOS technology [4]. This delay constitutes a constant speed limitation if a global signal is to be transmitted through the whole chip, not to mention the whole board.

## 1.2.2. Design Complexity and Layout Factors

Increased circuit complexity has been addressed by the development of CAD tools to aid the designer. These tools often allow the designer to specify the functionality of a system at a structural level and have a computer program generate the actual chip layout. System architectures may vary from systolic arrays to random control logic, making it difficult to automatically design a clock distribution network that will function at high speeds for all possible structures. The routing of clock wires as well as the load on the clock signal are global considerations which are difficult to extract from local connectivity. A modular design approach as facilitated by asynchronous techniques greatly simplifies the design efforts of complicated systems and fits well with the macro-cell design methodology often adopted by ASIC CAD tools.

We mentioned that a constant delay is necessary to charge a large capacitance in CMOS technology. In synchronous design, this delay must be accommodated by a longer cycle time for any global signals if there are feedback connections. The considerations in layout and routing to reduce the effects of this delay demand expertise and design efforts. High-performance CMOS synchronous chips that achieve a 125 Mhz cycle time in simulation has been reported [30], but such design requires clever clock buffer schemes and usually results in a much higher design cost than a design that does not have any global constraints. Furthermore, if pipelining is used, asynchronous design allows capacitance delay to be divided and hidden into multiple pipeline stages without reducing the overall throughput.

### 1.2.3. Applicability to Board Level System Design

One of the primary advantages of using asynchronous design in implementing high performance digital systems is the ease of design at the board level, where clock distribution is not an issue. Inter-board communications have long used asynchronous links (for example the VME bus protocol) and inter-chip intra-board communications are becoming asynchronous too (for example the MC68000 series). Actually, in system design, it is not a question of whether to use asynchronous circuitry; the design problem is how low a level it is employed. At the network level, computing systems essentially never have global synchronization (each SUN workstation has its own clock). At the backplane level it is a mix; the higher performance backplanes are mostly asynchronous (for example VME and Multibus), but the lower performance personal computer backplanes are synchronous. Internally to a board, generally synchronous techniques are employed, but as the speeds increase and the density of the circuitry becomes higher, asynchronous interconnect of critical modules is desirable (as we see in the 68000 asynchronous I/O bus).

From a system design point of view, asynchrony means design modularity. As digital systems become complex, it is advantageous to adopt a modular design approach to simplifying the design task to only local considerations.

### 1.2.4. Programmable Architecture Considerations

The instruction cycle time for a pipelined architecture is determined by the longest processing delay among all pipeline stages. But this worst case is time-varying if an asynchronous processor is used, meaning that at different times (different instructions), this longest delay has different values. We can take advantage of this instruction-dependent processing time to obtain an *average* cycle time as opposed to the constant worst case cycle time exhibited by a synchronous processor. For example, a multiplication takes more time to compute than most other operations. If the number of multiplication instructions in a program is only a small fraction of the total number of instructions, the program can be completed within a much shorter period of time if variable processing times are allowed. Synchronous processors attempt to utilize variable processing times by subdividing the master clock into multiple phases, but the number of phases that can be derived from the master clock is severely limited by the necessity of non-overlap times between phases [31,32]. Asynchronous processors can be viewed as synchronous processors with a continuum of clock phases, without the provision in non-overlap times. Simulations of the operation of an asynchronous processor with some simple signal processing programs confirmed that a speed-up by a factor on the order of two or more can be expected due to this averaging effect. Chapter 4 will discuss this issue in more detail.

## 1.3. PLAN OF THE BOOK

This book concentrates on the synthesis and design of synchronization circuits in digital systems. Chapter 2, contributed by David G. Messerschmitt, describes synchronization and interconnection at an intuitive level, covering some basics required for the understanding of later chapters. Synchronization alternatives and a taxonomy of synchronization techniques will be introduced in this chapter as a pointer to later material. Chapter 3 gives a logic synthesis algorithm based on graph theory and Boolean algebra, from which a deterministic procedure is constructed for synthesizing self-timed circuits using anisochronous interconnection. In Chapter 4 the design of a self-timed programmable processor is discussed, based on the synthesis program developed in Chapter 3, and the design issues such as pipelining, data flow control, program flow control, initialization and feedback, I/O interface, and processor architectures will be addressed. In Chapter 5 a chip set designed to implement a vectorized adaptive lattice filter using asynchronous techniques will be described and the test results of these chips will be given. Chapter 6, contributed by David G. Messerschmitt, discusses the various isochronous interconnection schemes introduced in Chapter 2 in greater detail, along with their theoretical bounds on the throughputs. Chapter 7 covers an automatic verifier, contributed by Steve M. Nowick and David L. Dill, that has been used to verify various synthesized synchronization circuits and applied to the design a correct arbiter.

## REFERENCES

1. S. Y. Kung and R. J. Gal-Ezer, "Synchronous versus Asynchronous Computation in VLSI Array Processors," *SPIE, Real Time Signal Processing V* **341**(1982).

2. M. Hatamian and G. L. Cash, "Parallel Bit-Level Pipelined VLSI Designs for High-Speed Signal Processing," *Proceedings of the IEEE* **75**(9) pp. 1192-1202 ().

3.  D. F. Wann and M. A. Franklin, "Asynchronous and Clocked Control Structures for VLSI Based Interconnection Network," *IEEE Trans. on Computers* C-32(2)(March 1983).

4.  G. M. Jacobs and R. W. Brodersen, "Self-Timed Integrated Circuits for Digital Signal Processing Applications," *VLSI Signal Processing III*, IEEE PRESS, (November, 1988).

5.  T. H.-Y. Meng, R. W. Brodersen, and D. G. Messerschmitt, "A Clock-Free Chip Set for High Sampling Rate Adaptive Filters," *Journal of VLSI Signal Processing* 1 pp. 365-384 (April 1990).

6.  W. A. Clark, "Macromodular Computer Systems," *Proc. 1967 Spring Jt. Comp. Conf.*, pp. 335-336 Thompson Book Co., (1967).

7.  R. J. Swan, S. H. Fuller, and D. P. Siewiorek, "Cm*--A Modular Multiprocessor," *Proc. AFIPS 1977 Nat. Comput. Conf.*, pp. 637-644 (1977).

8.  S. H. Unger, *Asynchronous Sequential Switching Circuits*, Wiley-Interscience, John Wiley & Sons, Inc., New York (1969).

9.  D. A. Huffman, "The Synthesis of Sequential Switching Circuits," *J. Franklin Institutes* 257 pp. 161-190, 275-203 (March and April 1954).

10. R. E. Miller, *Switching Theory*, John Wiley & Sons, Inc., New York (1965).

11. G. Mago, "Realization Methods for Asynchronous Sequential Circuits," *IEEE Trans. on Computers* C-20(3)(March 1971).

12. S. H. Unger, "Asynchronous Sequential Switching Circuits with Unrestricted Input Changes," *Trans. on Computers* C-20(12) pp. 1437-1444 (Dec. 1971).

13. D. B. Armstrong, A. D. Friedman, and P. R. Manon, "Design of Asynchronous Circuits Assuming Unbounded Gate Delays," *IEEE Trans. on Computers* C-18(12)(Dec. 1969).

14. A. S. Wojcik and K.-Y Fang, "On the Design of Three-Valued Asynchronous Modules," *IEEE Trans. on Computers* C-29(10)(Oct. 1980).

15. J. C. Jr. Sims and H. J. Gray, "Design Criteria for autoasynchronous Circuits," *Proc. of Eastern Joint Computer Conf.*, pp. 94-99 (Dec. 1958).

16. D. E. Muller, "Asynchronous Logics and Applications to Information Processing," *Proc. Symp Applications of Switching Theory in Space Tech.*, pp. 289-297 Stanford University Press, (1963).

17. D. Hammel, "Ideas on Asynchronous Feedback Networks," *Proc. 5th Ann. Symp. on Switching Circuit Theory and Logic Design*, pp. 4-11 (Nov. 1964).

18. G. R. Couranz and D. F. Wann, "Theoretical and Experimental Behavior of Synchronizers Operating in the Metastable Region," *IEEE Trans. on Computers* C-24(6)(June 1975).

19. W. Plummer, "Asynchronous Arbiters," *IEEE Trans. on Computers* C-21(1)(Jan. 1972).

20. D. Kinniment and J. Woods, "Synchronization and Arbitration Circuits in Digital Systems," *Proc. IEE (England)* 123(10)(1976).

21.  J. Calvo, J. I. Acha, and M. Valencia, "Asynchronous Modular Arbiter," *IEEE Trans. on Computers* C-35(1)(Jan. 1986).

22.  J. C. Barros and B. W. Johnson, "Equivalence of the Arbiter, the Synchronizer, the Latch, and the Inertial Delay," *IEEE Trans. on Computers* C-32(7)(July 1983).

23.  J. Hohl, W. Larsen, and L. Schooley, "Prediction of Error Probability for Integrated Digital Synchronizers," *IEEE J. on Solid-State Circuits* SC-19(2)(1982).

24.  D. L. Dill and E. M. Clarke, "Automatic Verification of Asynchronous Circuits Using Temporal Logic," *Proc. 1985 Chapel Hill Conference on VLSI*, pp. 127-143 (1985).

25.  T. E. Williams, M. Horowitz, R. L. Alverson, and T. S. Yang, "A Self-Timed Chip for Division," *Advanced Research in VLSI, Proc of 1987 Stanford Conference*, pp. 75-96 (March 1987).

26.  S. E. Schuster, et al., "A 15ns CMOS 64K RAM," *1986 IEEE ISSCC Digest of Techinal Papers*, pp. 206-207 (February 1986).

27.  K.C. Saraswat and F. Mohammadi, "Effect of Scaling of Interconnections on the Time Delay of VLSI Circuits," *IEEE Journal of Solid State Circuits* SC-17(2)(April 1982).

28.  H. B. Bakoglu, "Circuit and System Performance Limits on ULSI: Interconnections and Packaging," *Ph.D. Dissertation, EE Department, Stanford University No. G541-4*, (October 1986).

29.  Y. Pauleau, "Interconnect Materials for VLSI Circuits - Part III," *Solid State Technology*, pp. 101-105 (June 1987).

30.  K. K. Parhi and M. Hatamian, "A High Sampling Rate Recursive Digital Filter Chip," *VLSI Signal Processing III*, IEEE PRESS, (November, 1988).

31.  Texas Instruments, *Details on Signal Processing: TMS320C30*, (September 1989).

32.  Motorola, *DSP56000 Digital Signal Processor User's Manual*, (1986).

1. Calvo, R. A., Ceccato, and M. Ulbrich, "An Overview of Modular Neural..."
   IEEE's Transactions on Computers, C-28 (Jan. 1989).

22. McClamoch, and B. W., "The Road Right Issue of the Number, the Knowledge
   ...ing, the Logan and the central delay, 1872, Trang, Au, Computers."
   J.D. (Feb. 1989).

3. ...d. Thoel, W. Laan, and L. Schooley, "The Sense of The Possibility for
   Antennas, Oига of Spatio-motion," IEEE J. ...o, Solid-State Circuits, 28
   ...4 (C 1882)

4. ...ca, B. W., and W. ...bach, "Appearance Value, ...ism of Harmonious Off-
   ...ami Using Templates, ...ca, ...ono, ...d, IEEE J. ...olid State, ...on 1874,
   pp. 23-43 4093 in

22. T. ...b ..., ...a ..., L. Altos, K. L., Thorsten, L. J. P. E. Yang, "...id. Cross-
   C. Raj Laszlo, ...tic Memory. Fractology ...icit Systems, J. IEEE, Standard
   Coding ..., pp. 72 ...a ..., (... 1874).

5. ...t. L. Salocta ..., "...O, ...O, ...O, ...M. T.O...ISSN #2., pp 1-
   ...ram Solr ..., No. 10, 1174 sub-o ...

...
...O ..., ...to ..., ...e-...o ...on ... 20 ... ... ...y ...o ...e ...on
...O, J. Yang, 2.8L ...c ...ment ..., 1274...
...ISK ... 1, 1-412

B. C. Searaj, J. C., ...d ..., ...NY, ...achter ...,...o, ...W. Mump
...tector and Classify, 1881, ...nce on, EE Department. Stanford
University N. Oxford ...hni...(...1889).

9. ...A. Weiss, "Theory...nd ...ance, Altos for W.U.S. sutut, ...U.S 1869, ...ap.
   Technology, pp. 1872...o, ...ant...

10. ...K. N. Polk and M. ...mann, "On P., S.ti-...o, ...etic, Purpose, I., ...c, 1,
   ...D, ...on-TN, ...perR, ...an-E-...c, ...o, ...e, ...ico...,...1 ...8.

11. ...c ..., L. Part ..., 190, ...o, ...ime Value, Data, ...ing, ...e Co, Program...
   1869

12. Motorola, ...786 000 ...ani, S-6717, ...icroo ...ing, Machine, 196.

# 2

# SYNCHRONIZATION

David G. Messerschmitt
Department of Electrical Engineering and Computer Sciences
University of California at Berkeley

In typical digital systems, there are many operations that are performed, some of which must proceed sequentially (observe a data precedence) and others of which can proceed concurrently. Correct system operation requires that these operations be properly coordinated. The role of synchronization in the system is to insure that operations occur in the logically correct order, and is critical in insuring correct and reliable system operation. As the physical size of a system increases, or as the speed of operation increases, synchronization plays an increasingly dominant role in the system design.

Digital communication has developed a number of techniques to deal with synchronization on a global and even cosmic scale, and as the clock speeds of chip, board, and room-sized digital systems increase they may benefit from similar techniques. Yet, the digital system and digital communication communities have evolved synchronization techniques independently, choosing different techniques and different terminology. We will now attempt to present a unified framework and terminology for synchronization design in digital systems, borrowing techniques and terminologies from both digital system and digital communication design disciplines [1,2]. We also attempt to define a taxonomy of different synchronization approaches that have been used or proposed in digital systems, again borrowing terminology from digital communication. Subsequently, in chapter 6, a number of alternative synchronization schemes are introduced, including the very popular synchronous interconnection, and their performance is compared to self-timed logic.

## 2.1. SYNCHRONIZATION IN DIGITAL SYSTEMS

Operations in digital systems can either proceed concurrently, or they must obey a precedence relationship. If two operations obey a precedence, then the role of synchronization is to insure that the operations follow in the correct order. Synchronization is thus a critical part of digital system design.

The most common approach to synchronization is to distribute a clock signal to all modules of the system, and then latch the input signals to these modules using the transitions of the clock. This approach has been used very successfully, to the extent that it is almost ubiquitous in past designs. However, with the scaling of feature-sizes in VLSI design, clock speeds are increasing rapidly, but increases in complexity tend to prevent significant reductions in chip size. As a consequence of this scaling, clock speeds in digital system designs are increasing in relation to propagation delays. This is causing increasing problems with the traditional synchronous design methodologies, certainly at the system and board levels, and increasingly even within high performance chips [3]. This problem will be accentuated with the more common application of optics to system interconnection.

Problems such as large propagation delay have been faced since the earliest days of digital communication system design, and hence there are a number of opportunities to apply digital communication principles to VLSI and digital system design. However, because of the wide difference between system physical sizes in relation to clock speeds, the design styles of the communities have developed almost independently.

In this chapter, we place the synchronization problem and design approaches in digital communication and digital system design in a common framework. In attaining the unification of design methodologies that we attempt in this chapter, the first difficulty we face is the inconsistencies, and even contradictions, between terms as used in the two fields. For example, the term "self-timed" generally means "no independent clock", but as used in digital communication it means no clock at all (in the sense that if a clock is needed it can be derived from the data), and in digital system design it indicates that there is a clock signal that is slaved to the data rather than being independent. Therefore, in section 2 we attempt to define a taxonomy of terminology that can apply to both fields, while retaining as much of the terminology as presently used as possible. We discuss this terminology in terms of the necessary levels of abstraction in the design process.

While asynchronous interconnect shows promise for the future, past practice is to use synchronous interconnect, as described further in chapter 6. In chapter 6 some fundamental limits on the throughput of synchronization techniques are derived, and synchronous and asynchronous techniques are compared in these terms. Chapter 6 also discusses how some alternative synchronization techniques from digital communication might be beneficial in digital system design, particularly in reducing dependence of throughput on interconnect delay.

## 2.2. ABSTRACTION IN SYNCHRONIZATION

A basic approach in system design is to define *abstractions* that enable the designer to ignore unnecessary details and focus on the essential features of the design. While every system is ultimately dependent on underlying physical laws, it is clear that if we relied on the solution of Maxwell's equations at every phase of the design, systems could never get very complex.

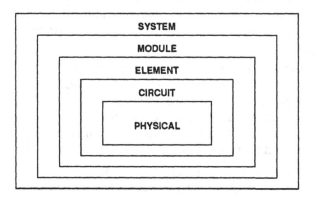

**Figure 2-1.** An example of some abstractions applied to digital system design.

Abstractions are often applied in a hierarchical fashion, where each layer of abstraction relies on the essential features of the abstraction level below, and hides unessential details from the higher level. This as illustrated for digital system design in figure 2-1. At the base of the design, we have the physical representation, where we have semiconductor materials, interconnect metalization, etc., and we are very concerned about the underlying physical laws that govern their properties. Above this, we have the circuit abstractions, where we deal with circuit entities such as transistors, interconnections, etc. Our descriptions of the circuit entities attempt to ignore the details of the physical laws underlying them, but rather characterize them in terms that emphasize their operational properties, such as current-voltage curves, transfer functions, etc. Above the circuit abstraction is the element, which groups circuit entities to yield slightly higher functions, such as flip-flops and gates. We describe elements in terms that deal with their operational characteristics (timing diagrams, Boolean functions, etc.) but ignore the details of how they are implemented. Above the element abstraction, we have the module (sometimes called "macrocell") abstraction, where elements are grouped together to form more complex entities (such as memories, register files, arithmetic-logic units, etc.) Finally, we group these modules together to form systems (like microcomputers).

Many other systems of abstractions are possible, depending on circumstances. For example, in digital communication system protocols, we extend the abstractions in figure 2-1, which carry us to the extreme of hardware design, to add various abstractions that hierarchically model the logical (usually software-defined) operation of the system (including many layers of protocols). Similarly, the computer industry defines other system abstractions such as instruction sets, operating system layers, etc.

In the present chapter, we are concerned specifically with synchronization in the design of digital systems and digital communication. In the context of digital systems, by *synchronization* we mean the set of techniques used to insure that operations are performed in the proper order. The following subsections define some appropriate abstractions for the synchronization design. This systematic treatment gives us a common base of terminology for synchronization design, and is utilized in the following subsections.

While abstractions are very useful, and in fact absolutely necessary, they should be applied with care. The essence of an abstraction is that we are ignoring some details of the underlying behavior, which we hope are irrelevant to the operation, while emphasizing others that are most critical to the operation. It should always be verified that the characteristics being ignored are in fact irrelevant, and with considerable margin, or else the final design may turn out to be inoperative or unreliable. This is especially true of synchronization, which is one of the most frequent causes of unreliable operation of a system. In the following, we therefore highlight behaviors that are ignored or hidden by the abstraction.

## 2.2.1. Boolean Signals

A *Boolean signal* (voltage or current) is assumed to represent, at each time, one of two possible levels. At the physical level, this signal is generated by saturating circuits and bistable memory elements. There are a couple of underlying behaviors that are deliberately ignored in this abstraction: the *finite rise-time* and the *metastable* behavior of memory elements.

Finite rise-time behavior is illustrated in figure 2-2 for a simple *RC* time constant. The deleterious effects of rise-time can often be bypassed by the *sampling* of the signal. In digital systems this is often accomplished using edge-triggered memory elements. In digital communication, rise-time effects are often much more severe because of the long distances traversed, and manifest themselves in the more complex phenomenon of *intersymbol interference*. In this case, one of several forms of equalization can precede the sampling operation.

Metastability is an anomalous behavior of all bistable devices, in which the device gets stuck in an unstable equilibrium midway between the two states for an indeterminate period of time [5]. Metastability is usually associated with the sampling (using an edge-triggered bistable device) of a signal whose Boolean state can change at any time, with the result that sampling at some point very near the transition will occasionally occur. Metastability is less severe a problem than rise-time in the sense that it happens only occasionally, but more severe in that the condition will persist for an indeterminate time (like a rise-time which is random in duration).

The Boolean abstraction is most valid when the rise-time is very much shorter than the interval between transitions, and metastability is avoided or carefully controlled. But the designer must be sure that these effects are negligible, or unreliable system operation will result.

00110001111110111001100100110...

**Figure 2-2.** Illustration of a digital signal with a finite rise-time as generated by an *RC* time-constant, where the Boolean signal abstraction is shown below.

## 2.2.2. Signal Transitions

For purposes of synchronization we are often less concerned with the signal level than with the times at which the signal changes state. In digital systems this time of change would be called an *edge* or *transition* of the signal. The notion of the transition time ignores rise-time effects. In fact, the transition time is subject to interpretation. For example, if we define the transition time as the instant that the waveform crosses some slicer level (using terminology from digital communication), then it will depend on the slicer level as illustrated in figure 2-3. Even this definition will fail if the waveform is not monotonic in the region of the slicer level.

Transition times are a useful abstraction for the case where the rise times are very short in relation to the interval between transitions, with the result that the variation in the transition time is negligibly small over the set of all possible definitions of the transition time. Rise-time is governed by underlying physical phenomena, such as transmission line dispersion, and can be reduced by using wider bandwidth drivers or intermediate repeaters. As system clock rates increase, however, for a given interconnect style the behavior ignored by this abstraction inevitably becomes important.

The transition abstraction can be extended to the notion of *uniformly spaced*

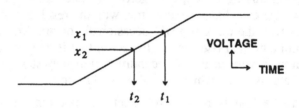

**Figure 2-3.** The transition time $t_i$ depends on slicer level $x_i$.

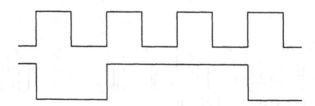

**Figure 2-4.** A Boolean signal (below) and associated clock (above). The transitions of the Boolean signal are slaved to the positive transitions of the associated clock.

*transitions*. For example, a clock signal can be modeled as a square wave, in which the transitions alternate in sign, and each adjacent pair of transitions (called a *cycle*) represents a time equal to the reciprocal of the clock *frequency*. For a data signal, transitions may or may not be present depending on the Boolean data (see figure 2-2), so we have to introduce the notion of the times where transitions *might* occur, called a *transition opportunity*, whether they actually occur or not. A data signal whose transitions are slaved to a clock with uniformly spaced transitions then has uniformly spaced transition opportunities (an example is shown in figure 2-4). We can think of these transitions as being associated with a clock that has positive transitions at identical times, which we call the *associated clock*, whether or not such a clock signal exists physically.

Uniformly spaced transitions ignore possible jitter effects in the generation or transmission of the Boolean signals, which often result in small variations in the times between transitions. Hence the need to define the concepts of instantaneous phase and frequency.

### 2.2.3. Phase and Frequency

For a Boolean signal, we can define a phase and frequency of the signal as the phase and frequency of the associated clock. It is convenient to describe mathematically a clock signal with uniformly spaced transitions as

$$x(t) = p((ft + \phi) \text{ modulo } 1), \qquad (2.1)$$

where $p(t)$ is a 50% duty cycle pulse

$$p(t) = \begin{cases} 1, & 0 \le t < 0.5 \\ 0, & 0.5 \le t \le 1 \end{cases}, \qquad (2.2)$$

$f$ is the *nominal frequency*, and $\phi$ is the *phase*. As $\phi$ varies over the range $0 \le \phi < 1$, the transitions are shifted in time over one cycle. The phase is thus expressed as the fraction of a cycle. When we have two Boolean signals, the *relative phase* can be expressed as the phase difference $(\phi_1 - \phi_2)$ between their respective associated clocks.

A more general model that includes more possible effects replaces (2.1) by

$$x(t) = p(((f + \Delta f)t + \phi(t)) \text{ modulo } 1), \qquad (2.3)$$

where $f$ is the *nominal frequency* of the associated clock, $\Delta f$ is a possible *offset* in the nominal frequency, and $\phi(t)$ is the *instantaneous phase variation* vs. time. The intention here is that $\phi(t)$ doesn't embody a frequency offset, but rather any offset from the nominal frequency is summarized by $\Delta f$. The precise mathematical conditions for this are complicated by the modulo operation, and also depend on the model for $\phi(t)$ (deterministic signal, random process, etc.). For example, if $\phi(t)$ is assumed to be a deterministic differentiable and continuous function (with no phase jumps) then it suffices for $\phi(t)$ to be bounded,

$$\phi(t) \le \phi_{max} , \qquad (2.4)$$

and for such a function the derivative (instantaneous frequency) must average to zero,

$$\overline{\frac{d\phi(t)}{dt}} = 0 \qquad (2.5)$$

(where the average is interpreted as a time average).

The model of (2.3) makes the assumption that the average frequency is a constant, although that average frequency may not be known *a priori* (for example when it depends on the free-running frequency of an oscillator). Such a signal is said to be *isochronous* (from "iso", the Greek root for "equal"), whereas if the frequency is not constant ($\Delta f$ is actually a function of time) the signal is said to be *anisochronous* (or "not equal"). An anisochronous signal can be modeled using (2.3), but the resulting phase will not be bounded. Thus, the essential difference between isochronous and anisochronous signals is the bounded phase condition of (2.4).

The time-varying phase in (2.3) is crucial where we cannot ignore small variations in the intervals between transitions, known as *phase jitter*. This jitter is usually ignored in digital system design, but becomes quite significant in digital communication, especially where the Boolean signal is passed through a chain of regenerative repeaters [4].

Directly following from the concept of phase is the *instantaneous frequency*, defined as the derivative of the instantaneous phase,

$$f(t) = f + \Delta f + \frac{d\phi(t)}{dt} . \qquad (2.6)$$

From (2.5), the *instantaneous frequency deviation* $d\phi(t)/dt$ has mean value zero, so that $(f + \Delta f)$ is the *average frequency*.

Given two signals $x_i(t)$, $i = 1,2$ with the same nominal frequency, and frequency and phase offsets $\Delta f_i$ and $\phi_i(t)$, the *instantaneous phase difference* between the two signals is

$$\Delta\phi(t) = (\Delta f_1 - \Delta f_2)t + (\phi_1(t) - \phi_2(t)) . \qquad (2.7)$$

## 2.2.4. Synchronism

Having carefully defined some terms, we can now define some terminology related to the *synchronization* of two signals. A taxonomy of synchronization possibilities is indicated in figure 2-5 [6,4]. As previously mentioned, a Boolean signal can be either isochronous or anisochronous. Given two

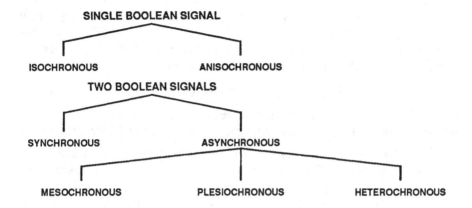

**Figure 2-5.** A taxonomy of synchronization approaches.

Boolean signals, if both are isochronous, the frequency offsets are the same, and the instantaneous phase difference is zero,

$$\Delta\phi(t) = 0 \qquad (2.8)$$

then they are said to be *synchronous* (from the Greek "syn" for "together"). Common examples would be a Boolean signal that is synchronous with its associated clock (by definition), or two signals slaved to the same clock at their point of generation. Any two signals that are not synchronous are *asynchronous* (or "not together"). Some people would relax the definition of synchronous signals to allow a non-zero phase difference that is *constant* and *known*.

In practice we have to deal with several distinct forms of asynchrony. Any signal that is anisochronous will be asynchronous to any other signals, except in the special and unusual case that two anisochronous signals have identical transitions (this can happen if they are co-generated by the same circuit). Thus, anisochrony is a form of asynchrony.

If two isochronous signals have exactly the same average frequency $f + \Delta f$, then they are called *mesochronous* (from the Greek "meso" for "middle"). For mesochronous signals, the fact that each of their phases is bounded per (2.4) insures also that the phase difference is also bounded,

$$\Delta\phi(t) \le 2\phi_{max} , \qquad (2.9)$$

a fact of considerable practical significance. Two signals generated from the same clock (even different phases of the same clock), but suffering indeterminate interconnect delays relative to one another, are mesochronous.

Two signals that have average frequencies that are nominally the same, but not exactly the same (usually because they are derived from independent oscillators), are *plesiochronous* (from the Greek "plesio" for "near"). Suppose the nominal frequencies are both $f$, but the actual frequencies are $f + \Delta f_1$ and $f + \Delta f_2$, then the instantaneous phase difference is

$$\Delta \phi(t) = (\Delta f_1 - \Delta f_2)t + (\phi_1(t) - \phi_2(t)) , \qquad (2.10)$$

where the first term increases (or decreases) linearly with time. For two plesiochronous signals, it can't be predicted which has the higher frequency.

Finally, if two signals have *nominally different* average frequencies, they are called *heterochronous* (from the Greek "hetero" for "different"). Usually the tolerances on frequency are chosen to guarantee that one signal will have an actual frequency guaranteed higher than the other (naturally the one with the higher nominal rate). For example, if they have nominal frequencies $f_1$ and $f_2$, where $f_1 < f_2$, and the worst-case frequency offsets are $|\Delta f_1| \leq \eta_1$ and $|\Delta f_2| \leq \eta_2$, then we would guarantee this condition if $f_1$ and $f_2$ were chosen such that

$$f_1 + \eta_1 < f_2 - \eta_2 . \qquad (2.11)$$

## 2.3. TIMING ABSTRACTION IN DIGITAL SYSTEMS

The previous section has covered some general considerations in the modeling of Boolean signals from a synchronization perspective. In this section we describe a couple of additional abstractions that are sometimes applied in digital system design, and represent simplifications due to the relatively small physical size of such systems.

### 2.3.1. Equipotential Region

Returning to figure 2-1, it is often assumed at the level of the element abstraction that the signal is identical at all points along a given wire. The largest region for which this is true is called by Seitz [7] the *equipotential region*. Like our other models, this is never strictly valid, but is useful if the actual time it takes to equalize the potential along a wire is small in relation to other aspects of the signals, such as the associated clock period or risetime. The equipotential region is a useful concept in digital system design because of the relatively small size of such systems. However, the element dimensions for which it is valid is decreasing because of increases in clock frequency with scaling, and a single chip can generally no longer be considered an equipotential region.

## 2.3.2. Ordering of Signals

In the design of digital systems it is often true that one Boolean signal is *slaved* to another, so that at the point of generation one signal transition can always be guaranteed to precede the other. Conversely, the correct operation of circuits is often dependent on the correct ordering of signal transitions, and quantitative measures such as the minimum time between transitions. One of the main reasons for defining the equipotential region is that if a given pair of signals obey an ordering condition at one point in a system, then that ordering will be guaranteed anywhere within the equipotential region.

## REFERENCES

1.   D. G. Messerschmitt, "Synchronization in Digital Systems Design," *IEEE Trans. on Special Areas in Communications* **JSAC-8**(10)(October 1990).

2.   D. G. Messerschmitt, "Digital Communication in VLSI Design," *Proc. Twenty-Third Asilomar Conference on Signals, Systems, Computers*, (Oct. 1989).

3.   M. Hatamian, L. A. Hornak, T. E. Little, S. K. Tewksbury, and P. Franzon, "Fundamental Interconnection Issues," *AT&T Technical Journal* **66**(July 1987).

4.   E. A. Lee and D. G. Messerschmitt, *Digital Communication,* Kluwer Academic Press, Boston (1988).

5.   T. J. Chaney and C. E. Molnar, "Anomalous Behavior of Synchronizer and Arbiter Circuits," *IEEE Trans. on Computers* **C-22**(4)(April 1973).

6.   G. H. Bennett, "Pulse Code Modulation and Digital Transmission," *Marconi Instruments*, (April 1978).

7.   C. Mead and L. Conway, *Chap. 7, Introduction to VLSI Systems*, Addison-Wesley, Reading, Mass. (1980).

# 3

# SYNTHESIS OF SELF-TIMED CIRCUITS

This chapter describes a synthesis procedure of designing digital systems that do not require the distribution of a clocking signal. Guarded commands [1] were chosen to provide a simple syntax for circuit specifications. The notation of signal transition graphs [2] was used to represent circuit behavior and to simplify algorithms and graph manipulations. The definition of semi-modularity [3] was modified and the theorems developed for marked directed graphs [4] were used to guarantee a hazard-free implementation. All together, a deterministic algorithm of synthesizing self-timed synchronization circuits from high-level specifications was constructed. The implication is that fully asynchronous design (or more precisely, asynchronous design using anisochronous interconnect according to the previous chapter) is feasible and self-timed circuit synthesis can be automated. Although in this chapter we are primarily concerned with the synthesis of non-metastable circuits, the procedure is also valid for metastable circuit synthesis.

## 3.1. AN INTERCONNECTION SCHEME

The most common design approach in the past has been to distribute a clock signal throughout a digital system, and use this signal to insure synchronous operation. This is called *synchronous interconnect* and is discussed (with variations) in chapter 6.

In this section, we introduce a specific method of synchronization of aniso-chronous signals with ordered transitions. This approach, which is the subject of chapters 3-5, is commonly called *asynchronous handshaking* or *self-timed synchronization*. It uses extra handshaking leads to avoid the distribution of a clock signal, and the synchronization is accomplished by a four-phase handshake protocol. We will also introduce a logic family which supports this protocol.

There have been many excellent works on designing asynchronous handshaking circuits, often using the formal language syntax such as trace theory, Petri nets, temporal logic, regular expressions, pass expressions, etc. [5,6,7,8,9]. However, these formal languages are usually difficult to understand by the system designer, and difficult to use for specifying practical systems that tend to be much more complicated than the simple circuit examples given in the literature. Another concern is that it is often difficult to relate the circuits compiled from these formal languages with certain properties important to system design, for example the system throughput, hardware utilization, and overall system performance.

One approach to eliminating some of these problems is based on the synthesis of self-timed circuits from a high-level specification [10]. Self-timed circuits differ from the early asynchronous circuits in that they can be designed to be *speed-independent*; i.e. their behavior does not depend on any relative delays among physical gates [11,12,13]. A self-timed circuit is called *delay-insensitive* if, besides being speed-independent, its behavior does not depend on the wiring delays within a module (this module can be as large as a system or as small as a gate). As shown by Martin in [14], truly delay-independent circuits are of only limited use. Most of the self-timed circuits discussed in this book are speed-independent, but not insensitive to wiring delays (although they are insensitive to the wiring delays *between* modules). Since self-timed circuits guarantee correct timing by synthesis, timing simulations are required only to predict performance but not to verify correctness.

We are primarily concerned with synthesizing self-timed synchronization circuits with desired properties from a system point of view, as opposed to confining our attention to certain circuit modules. Therefore we start the discussion with an overview of the proposed design methodology.

A digital design is composed of two types of decoupled blocks: computation blocks and interconnection blocks (shown in figure 3-1). Computation blocks, which includes combinational logic such as shifters and multipliers and memory elements such as RAMs and ROMs, perform processor operations. Interconnection blocks, which include data transfer circuits such as pipeline registers and multiplexers (indicated in figure 3-1 by blocks filled with hashed lines), handle transfers of both control and data signals. A computation block generates a completion signal to indicate valid output and to request a data transfer. An interconnection block operates on these completion and request signals and insures that data are correctly transferred regardless of the relative delays among these handshake signals.

## 3.1.1. Computation Blocks

One type of computation blocks is combinational logic for memoryless function evaluation, and the other has memory elements for program control and data accessing. It has been shown that two binary handshake signals

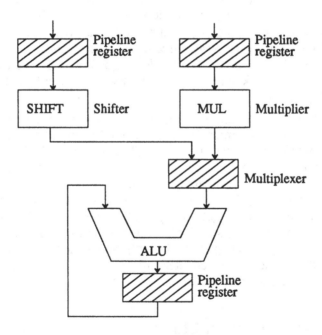

**Figure 3-1.** Using the block diagram design approach, a digital design can be specified by decoupled computation blocks and interconnection blocks.

(*request* and *complete* signals in our case) are necessary and sufficient to realize general asynchronous networks with unbounded gate delays [3]. For efficient hardware implementation, any combinational operation can be conveniently combined into one computation block, given that request and completion signals are generated along the datapath. Memory elements can be handled in the same way, as long as the memory access mechanism is initiated by a request signal and ended with a completion signal.

Our experimental implementations employ a logic family, differential cascode voltage switch logic (DCVSL) [15,16], to generate completion signals for combinational logic in a general way. A DCVSL computation block is shown in figure 3-2, where the *request* signal can be viewed, for the moment, as the completion signal from the preceding block. When *request* is low, the two output data lines would be pulled up by the p -MOS transistors and the *complete* signal will be low. When *request* goes high, which indicates that the computation of the preceding block has been completed and the differential input lines (both control and data) are stable and valid for evaluation, the two p -MOS transistors would be cut-off and the input lines will be evaluated by the NMOS tree. The NMOS logic tree is designed such that only one of the output data lines (two lines per data bit) will be pulled down by the NMOS tree once and only once, since there is no active pull-up device; then the *complete* signal will be set high by the NAND gate. Feedback transistors can be used at the output data lines to make DCVSL static.

Differential inputs and outputs are necessary for completion signal generation. Existing logic minimization algorithms [17] can be used to design differential NMOS logic trees, with a result that the hardware overhead is minimal compared with dynamic circuits. DCVSL has been found to offer a performance advantage compared with primitive NAND/NOR logic families, since NMOS logic trees are capable of computing complex Boolean functions within a single gate delay [18]. Several DCVSL computation blocks commonly used in signal processing, such as arithmetic-logic-units, shifters, and multipliers, have been designed and demonstrated a speed comparable to their synchronous counterparts [16,19]. The routing of differential input and output lines often requires on the order of a 40% active area penalty.

## 3.1.2. Interconnection Blocks

An interconnection block operates on the completion signals generated by computation blocks so that data can be properly transferred. Since requests for communication may be issued at any time, an interconnection circuit is ideally designed such that no erroneous behavior will occur under all temporal variations in circuit gate delays.

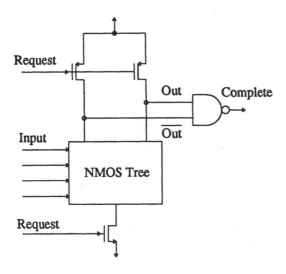

**Figure 3-2.** A schematic diagram of DCVSL for completion signal generation. The completion signal *Complete* goes high when output data lines become stable and stays low when *Request* is precharging the NMOS tree.

As an example, the simplest interconnection circuit is a pipeline handshake circuit. As shown in figure 3-3, two computation blocks labeled A and B are connected in a pipeline. Because of the pipeline handshake circuit, block A can process the next datum while block B processes the current one. Since block A might take longer to finish than block B or *vice versa*, an *acknowledge* signal is necessary to indicate when block B has completed its task and is ready for the next. The handshake circuit must prevent "runaway" conditions (data overwritten at the input to block B if block B has a long computation latency) or "continual feeding" (data computed more than once by block B if block A has a long latency). Other types of asynchronous handshake circuits display similar requirements.

A deterministic algorithm for interconnection circuit synthesis that meet these requirements will be described in this chapter. The algorithm systematically designs correct self-timed interconnection circuits with minimum constraints, and thus allows maximal possible concurrency. Based on the circuit model used, we can also assess certain optimality properties of self-timed circuits under very general conditions, and therefore provide a foundation for comparisons against other circuit approaches.

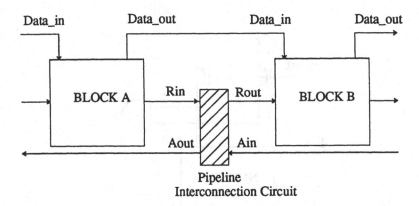

Figure 3-3. A simple example of an interconnection circuit: a pipeline interconnection circuit that controls data transfers between computation blocks A and B.

## 3.2. CIRCUIT BEHAVIORAL DESCRIPTION

We first give a brief overview of the history of asynchronous circuit design. Switching circuit theorems have been used to describe logic networks built of circuit elements with unbounded gate delays [3,12,20]. The first attempt to design asynchronous circuits for a real machine was for the arithmetic control unit of Illiac II using a flow chart approach [21,22,23]. Trace theory was later proposed for designing delay-insensitive circuits from programs which describe circuit trace structures [5], and Petri nets were investigated to model speed-independent circuit behavior [24,8,25,26,27]. The resulting circuits using these approaches tend to be complex [8]. Moreover, there is no control over the performance (as measured by throughput) because of the difficulty in optimizing the mapping or the decomposition process. The same problem arises in defining speed-independent modules that can be combined to form specific circuit functions [28,29,26,30].

Self-timed circuit *synthesis* has been an active research area in recent years [8,2,31,32,33,30], and the synthesis of self-timed control (including inter-connect) circuits using signal transition graphs [2] is a promising approach. Given a signal transition graph, the synthesis process modifies the graph to satisfy a number of syntactic requirements and derives a logic implementation from the modified graph. Nevertheless, the initial stage of constructing a signal transition graph from a desired synchronization scheme which guarantees the fastest operation is non-trivial, as will be illustrated by an example in the next section.

In the computation model as described in the previous section, computation circuits are decoupled from interconnection circuits. The computation blocks can be viewed as adding uncontrolled transmission delays between the terminals of an interconnection circuit. The synthesis procedure to be discussed in this chapter allows us to optimize the design of interconnection circuits as we perform the synthesis. Starting from a high-level specification, we describe a deterministic synthesis algorithm with no heuristics. The logic synthesized is guaranteed to be hazard-free and allows the operation among all the circuits that behave accordingly to the same specification.

### 3.2.1. Signal Transition Graphs

Classic asynchronous circuit design usually describes circuit behavior using a state diagram. A circuit resides in one of the states, and each state of a circuit is represented by a *bitvector* $(x_1, x_2, ..., x_n)$, where $x_1, x_2, ..., x_n$ are signals. The number of states ($\leq 2^n$) is finite. The behavior is defined by a set of sequences of states, where each sequence in the set is called an *allowed sequence*. To describe desired circuit behavior by a state diagram, all possible allowed sequences are enumerated. This approach has two disadvantages: first, the number of states grows exponentially with the number of signals; and second, it is often difficult to derive the correct state diagram which guarantees maximum concurrency. A simpler specification uses signal transitions instead of signal levels in the specification, which reduces the complexity of the input set from $2^n$ to $2n$.

The most common transition specification is the timing diagram used to describe circuit behavior of edge-triggered logic. A graphical equivalent of

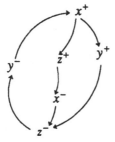

**Figure 3-4.** A signal transition graph where $x^-$ denotes the transition that $x$ goes from 1 to 0 and $x^+$ denotes the transition that $x$ goes from 0 to 1.

a timing diagram, called a *signal transition graph*, has been proposed as a general tool for specifying circuit behaviors [2]. In a signal transition graph, the rising transition of a signal $x$ is denoted as $x^+$ and a falling transition as $x^-$. For example, the signal transition graph shown in figure 3-4 specifies that the rising transition of signal $x$ ($x^+$) causes signals $y$ and $z$ to go high ($y^+$ and $z^+$), and that after $z$ has gone high ($s^+$), $x$ will go low ($x^-$), and so forth. An arc in a signal transition graph represents a *causal relation*; for example, $x^+ \rightarrow y^+$ means that $x$ must go high before $y$ can go high. We say that $x^+$ *enables* $y^+$ and $y^+$ is *enabled* by $x^+$. Signal transition graphs can be viewed as interpreted free-choice Petri nets [34], where the transitions in nets are named with signals and the places in nets are specified by causal relations. In the context of circuit synthesis, a free-choice Petri net can be decomposed into multiple marked directed graphs [4]. Marked directed graphs are a subclass of Petri nets in which each place has only one input transition and one output transition. Hence places can be replaced by simple arcs without ambiguity as in causal relations.

Signal transition graphs can be defined using marked directed graph notations [4]. A signal transition graph is a finite directed graph $G(V, A)$, where $V$ is the set of vertices (signal transitions) and $A$ is the set of arcs (causal relations). We assign a number of tokens to each arc, called a *marking* of the graph. A transition is said to be *enabled* if the number of tokens on every one of its incoming arcs is non-zero. For example, as shown in

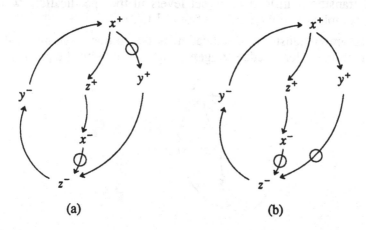

(a)                          (b)

**Figure 3-5.** (a). A signal transition graph with a token marking. Transition $y^+$ is enabled because its incoming arc is marked. (b). The token marking after the firing of $y^+$. The *firing* of an enabled transition consists of taking one token off each incoming arc and adding one token to each outgoing arc.

figure 3-5(a), transition $y^+$ is enabled as marked by the token on $x^+ \rightarrow y^+$. After the firing of $y^+$, the marking consists of the original token on $x^- \rightarrow z^-$ and a new token on $y^+ \rightarrow z^-$, as shown in figure 3-5(b).

Since a signal transition graph can be viewed as an interpreted marked directed graph, it is natural to analyze the circuit properties represented by a signal transition graph using the theorems for marked directed graphs. Even though these theorems do not address circuit synthesis directly, we can formulate them so as to assert certain circuit properties of interest. Some theorems even carry physical meaning for the corresponding circuit representation. The following are some relevant definitions and claims [4].

**Definition 3.1.** A marking is *live* if every transition can be enabled, or can be enabled through some sequence of firings.
**Claim 3.1.** A marking is live if and only if the token count of every simple loop of the graph is non-zero [4].

A live marking implies that every transition can be enabled after some firing sequence from the initial marking. It can be shown that a marking which is live remains live after firing [4]. Hence a live marking implies that a transition within a loop can be enabled infinitely often.

**Definition 3.2.** A marking is *safe* if no arc is assigned more than one token, and if no sequence of firings can bring two or more tokens to one arc.
**Claim 3.2.** A live marking is safe if and only if every arc in the graph is in a simple loop with token count one [4].

A safe marking insures that a graph has a state diagram representation. The markings shown in figure 3-5(a) and figure 3-5(b) are both live and safe since every arc in the graph is in a simple loop with token count one.

**Claim 3.3.** If the underlying undirected graph[1] of a marked directed graph G(V, A) is connected, then G can be assigned a live and safe marking if and only if G is strongly connected[2] [4].

Claim 3.3 states that if a marked directed graph has a live-safe marking, then the graph must be strongly connected. A strongly connected signal transition graph indicates constraints on input signal transitions, and can be used to describe the behavior of a *complete* circuit [3]. A complete circuit has no explicit input or output signals and represents a closed system where feedback from output signals to input signals is explicitly specified, as the

---

[1] The underlying undirected graph of a directed graph is the graph with directed arcs replaced by undirected arcs.

[2] A directed graph is strongly connected if every pair of nodes in the graph are mutually reachable from one another.

circuits shown in figure 3-5. It may not seem very interesting to consider a system without inputs and outputs, but in fact a complete circuit can be seen as a composition of partial open circuits with interwoven inputs and outputs. A complete circuit has more modeling power than an open circuit, because it is capable of specifying the inter-relations (called *environmental constraints*) between multiple open circuits using a graph notation.

> **Claim 3.4.** For every strongly connected marked directed graph, there exists a unique equivalence class of live-safe markings in which any two markings are mutually reachable from one another [4].

If each live-safe marking is assigned a state, then Claim 3.4 indicates the existence of a state graph, called a reachability graph, formed of mutually reachable live-safe markings. If the marked directed graph is a signal transition graph, we call the reachability graph the underlying *state diagram*.

From the above definitions and claims, safeness and liveness are properties of the initial state of a strongly connected graph rather than the properties of the graph structure. However, since every signal in a signal transition graph has at least two transitions (rising and falling), which represents a correlation between the transitions of the graph, we need a condition on the structure of the graph to insure that the underlying state diagram has a state assignment. This condition was observed in [2].

> **Claim 3.5.** Every state in the state diagram derived from a strongly connected signal transition graph can be assigned a bitvector if and only if every pair of signal transitions is within a simple loop [2].

An intuitive explanation of Claim 3.5 is as follows. If there is a pair of signal transitions, say $x^+$ and $x^-$, which is not within a simple loop of a strongly connected signal transition graph, then these two transitions can be enabled concurrently. The state in which both of the transitions are enabled does not have a definite signal level of $x$ in its bitvector, and therefore there is no bitvector state assignment for this state. We call a signal transition graph *live* if it satisfies the condition in Claim 3.5, because the same condition also implies that every transition in the circuit can be enabled infinitely often. For example, the signal transition graph shown in figure 3-5(a) is live, because it is strongly connected and every pair of signal transitions is within a simple loop. A marking represents a state in which a circuit resides; a live initial marking on a live signal transition graph guarantees that the circuit will not be deadlocked. From Claim 3.4, a reachability graph can be formed of mutually reachable live-safe markings and from Claim 3.5, every state in the reachability graph can be assigned a bitvector. In figure 3-5(a), if a state is defined to be $(x, y, z)$, then the state diagram (the reachability graph with state assignments) is shown in figure 3-6, where the live-safe marking shown in figure 3-5(a) corresponds to state $(0, 0, 1)$ and the live-safe marking shown in figure 3-5(b) corresponds to state $(0, 1, 1)$.

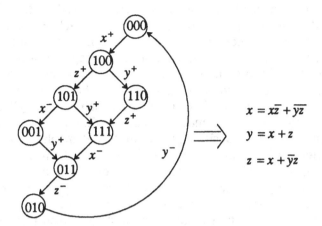

$$x = x\bar{z} + \overline{yz}$$

$$y = x + z$$

$$z = x + \bar{y}z$$

**Figure 3-6.** The state diagram with state assignments of bitvector $(x, y, z)$ and the corresponding Boolean expression for each signal. For example, the live-safe marking shown in figure 3-5(a) corresponds to state $(0, 0, 1)$ and the marking shown in figure 3-5(b) corresponds to state $(0, 1, 1)$.

## 3.2.2. Speed-Independent Circuits

In this subsection, we will briefly review the basic definition of speed-independent circuits following [3] and interpret the definition in terms of signal transition graphs. Moreover, the difference between a speed-independent circuit and a hazard-free implementation is discussed.

We now give the classic definition of speed-independent circuit, a definition modified from [3].

> **Definition 3.3.** A circuit is speed-independent if the state diagram representation of the circuit satisfies the following three conditions: (1) The state diagram is strongly connected. (2) Each state in the diagram can be assigned a unique bitvector. (3) If a signal transition is enabled in one state but not fired when the circuit changes into a second state, that signal transition must still be enabled in the second state.

This speed-independence definition is based on signal states, not on signal transitions. Hence the cardinality of the input set, the number of states, is exponential in the number of signals. The first condition insures that the circuit will not enter a final state; the second condition insures that the circuit is deterministic; and the third condition insures that if a transition takes longer to complete than other concurrent transitions, it will still be enabled before it is actually fired.

The sufficient conditions for each of the three speed-independence conditions on a signal transition graph can be easily derived, based on the fact that a circuit state diagram is the interpreted reachability graph of a signal transition graph. One sufficient condition for the first and third conditions to hold true is that the corresponding signal transition graph is strongly-connected and live. A live signal transition graph guarantees the existence of a state assignment for each marking (Claim 3.5) and the existence of a reachability graph (the state diagram) of mutually reachable states (Claim 3.4), thus satisfying the first speed-independence condition. The token marking mechanism on a signal transition graph asserts that enabling tokens will stay unconsumed on the enabling arcs until the enabled transition is actually fired, no matter which state (marking) the circuit might be in; hence, an enabled transition will remain enabled in the second state if not fired in the first state, satisfying the third speed-independence condition. Therefore, any live signal transition graph, if each state can be uniquely assigned, represents a speed-independent circuit by definition. For example, the circuit state diagram shown in figure 3-6, by definition, represents a speed-independent circuit.

A hazard-free circuit *implementation* is another issue. First we need a definition of a hazard [35]: If it is possible for an output signal to behave in a manner different from its predicted behavior, the circuit is said to contain a *hazard* for that input transition pattern.

From the discussion of speed-independent circuits, it would seem that they can be implemented using edge-triggered logic elements without hazards, as suggested by the fact that a live signal transition graph actually represents a speed-independent circuit under the unique state assignment assumption. However, edge-triggered logic is composed of level-triggered logic with timing constraints on set-up times and transition slopes. We wish to avoid timing constraints, because they are technology dependent. We therefore base our synthesis on basic level-triggered logic, whose behavior does not depend on signal timing, but solely on signal levels. To get a level-triggered logic implementation, Boolean algebra can be used to derive logic expressions from a state diagram which satisfies the three conditions in Definition 3.3. The state diagram shown in figure 3-6 is speed-independent and the corresponding Boolean expressions for the three signals $x$, $y$, and $z$ are also shown in the figure. In Section 3.4, we will shown that the circuit implemented by these three Boolean expressions is not hazard-free. In other words, the three conditions on speed-independent circuits describe ideal circuit *behavior*, but do not guarantee hazard-free *implementation*. In order to implement hazard-free speed-independent circuits using level-triggered logic elements, stronger conditions than just liveness and safeness must be devised.

## 3.3. INTERCONNECTION CIRCUIT SPECIFICATIONS

The simplest interconnection circuit, a pipelining handshake circuit shown in figure 3-3, checks the input request signal $R_{in}$ (the completion signal of computation block A) to see if the output datum of block A is valid, and checks the feedback acknowledge signal $A_{in}$ to see if block B is ready for a new input. $R_{out}$ controls the request signal to the DCVSL of block B, indicating when block B should start evaluation. $A_{out}$ controls the acknowledge signal to the interconnection block preceding block A, notifying block A when its output datum is transferred to block B.

The common four-phase handshake protocol works as follows [1,3,33,36]. Assume that the four signals $R_{in}$, $R_{out}$, $A_{in}$, and $A_{out}$ are initially at level 0 ($R_{in}^{-}$, $R_{out}^{-}$, $A_{in}^{-}$, $A_{out}^{-}$). When block A finishes its computation, it raises $R_{in}$ ($R_{in}^{+}$), the completion signal from the DCVSL of block A, to request for a data transfer to block B. Since $A_{in}$ is initially low, meaning that block B is ready to accept a new input, the handshake circuit raises $A_{out}$ ($A_{out}^{+}$) to tell block A that its output datum has been accepted so that $R_{in}$ can be reset ($R_{in}^{-}$). The handshake circuit then raises $R_{out}$ ($R_{out}^{+}$) to initiate the computation in block B. Eventually block B will complete its task and output a completion signal. This information is fed back through $A_{in}$ ($A_{in}^{+}$) so that $R_{out}$ can be reset. Resetting $R_{out}$ ($R_{out}^{-}$) will in turn reset $A_{in}$ ($A_{in}^{-}$) and complete the four-phase handshake loop. The four-phase handshake protocol always uses the rising transitions to initiate operation and the falling transitions to reset. The modes of operation coincide with that of DCVSL since a rising transition of request ($R_{out}$ of the handshake circuit) initiates evaluation and the falling transition of request starts the precharge. If DCVSL is to be used for completion signal generation, handshake circuits must follow this four-phase discipline. The two-phase handshake [36] is more efficient in the sense that only two transitions are needed for each operation cycle, but unfortunately we are not aware of a logic family that can generate completion signals in a two-phase cycle without using edge-triggered circuitry or a four-phase to two-phase converter.

In synthesizing the Boolean functions for $A_{out}$ and $R_{out}$, if too weak a condition is used to set or reset $A_{out}$ and $R_{out}$, samples might be flushed out before completion, or samples might be used twice in the same computation block (this problem was recognized in the design in [37]). If too strong a condition is used to insure proper operation, the handshake circuit may incur an unnecessarily long delay in response to a request, or suffer a low percentage of hardware utilization (this happens in [32]). The design of a reliable and fast handshake circuit is non-trivial.

The four-phase handshake protocol dictates that the sequence of signal transitions on the right hand side of a handshake circuit is always the iterative $R_{out}^+ \rightarrow A_{in}^+ \rightarrow R_{out}^- \rightarrow A_{in}^-$ and on the left hand side $R_{in}^+ \rightarrow A_{out}^+ \rightarrow R_{in}^- \rightarrow A_{out}^-$. The minimal requirement for a handshake circuit is that its transitions follow this sequence. To specify these assumed sequences for every possible connection in a signal transition graph shifts attention from circuit behavior, our primary concern, to internal details of signal transitions. More importantly, it makes it difficult to determine a specification which will preserve the desirable properties. We need a higher-level description of circuit behavior, and Dijkstra's guarded commands were chosen for this purpose [1,33,38].

A guarded command consists of a statement list prefixed by a Boolean expression: only when this Boolean expression is true, is the statement list eligible for execution. We use the subclass of deterministic guarded commands (table 3-1), since metastable circuits will not be considered in this chapter.

A guarded command set is formed by combining the commands in table 3-1 in a language syntax formally defined in [1]. AND commands are self-

| Deterministic Guarded Commands | |
|---|---|
| Notation | Interpretation |
| Basic guarded command: <br> $[C \rightarrow S]$ | $C$ is a *pre−condition* and $S$ is a list of statements to be executed if $C$ is true. |
| AND command: <br> $[C_1 \cap C_2 \cap \cdots \cap C_n \rightarrow S]$ | $C_i$ is a pre-condition and $S$ is to be executed if every $C_i$ is true. |
| OR command: <br> $[C_1 \cup C_2 \cup \cdots \cup C_n \rightarrow S]$ | $S$ is to be executed if any one of $C_i$ is true. But for the purpose of determinism, only one of $C_i$ can be true at a time. |
| Sequential Construct: <br> $[C_1 \rightarrow S_1; \; C_2 \rightarrow S_2]$ | $C_2$ can be tested only after $S_1$ has been executed. |
| Parallel Construct: <br> $[C_1 \rightarrow S_1 \parallel C_2 \rightarrow S_2]$ | The two clauses $C_1 \rightarrow S_1$ and $C_2 \rightarrow S_2$ can be processed concurrently. |
| Alternative Construct: <br> $[C_1 \rightarrow S_1 \vert C_2 \rightarrow S_2]$ | $S_i$ is executed if $C_i$ is true, but only one of the $C_i$ can be true at a time. |
| Repetitive construct: <br> $*[C \rightarrow S]$ | The clause $[C \rightarrow S]$ is to be repeatedly executed. |

**Table 3-1.** Legal constructs of deterministic guarded commands.

explanatory. Deterministic OR commands insure that the circuits are safe under multiple pre-conditions. A general OR command, which allows more than one pre-condition to be valid at a time, corresponds to the property of unsafeness in the Petri net theory, which is not allowed in our circuit specifications. Sequential constructs specify the precedence relation in circuit behavior, while parallel constructs specify the possible concurrency. Alternative constructs specify one of several alternative actions to be taken according to the pre-conditions. Although we require that only one of the pre-conditions can be true at a time (deterministic alternative constructs), the synthesis procedure described in this chapter is also valid for synthesizing general alternative constructs, in which more than one pre-condition can be valid. The resulting circuit behavior of a general alternative construct is not deterministic and may display metastability. Repetitive constructs are necessary since they guarantee the liveness condition as imposed by the synthesis algorithm.

Circuit specification is like programming, in that behavioral descriptions are translated into formulae to be operated upon. It is difficult to address the desirable properties for specifications, just as it is difficult to judge a programming style. But the performance resulting from different specifications is a substantive issue. What guarded commands provide us is a convenient bridge between operations expressed by block diagrams and signal transition graphs. Given a guarded command, our synthesis procedure guarantees an implementation which allows the fastest operation under the hazard-free conditions, if such an implementation exists. The correctness of a specification lies in the existence of a hazard-free implementation. The synthesis procedure works like an efficient compiler, which can generate the most efficient object code given a program but cannot *guess* what the programmer intends. Since our primary interest is to synthesize self-timed circuits from a simple high-level specification, guarded commands are found to be adequate for this purpose. We do not address properties such as *accuracy* or *equivalence* here, though these properties are interesting in their own right.

Two specifications relating the input and the output of a pipeline interconnect circuit using the basic four-phase handshake protocol are illustrated in figure 3-7 and figure 3-8. In figure 3-7, the guarded command specification is

$$* [R_{in}^+ \rightarrow R_{out}^+], \tag{3.1}$$

meaning that if the input data is ready ($R_{in}^+$), then start computation ($R_{out}^+$). The requirement that the succeeding block must be empty ($A_{in}^-$) before a new input can be accepted is reflected in the four-phase handshake transitions $A_{in}^- \rightarrow R_{out}^+$. In figure 3-8, the guarded command specification is

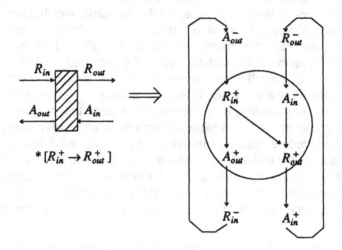

**Figure 3-7.** A pipeline interconnection circuit specified by $*[R_{in}^+ \to R_{out}^+]$ and its graphical representation.

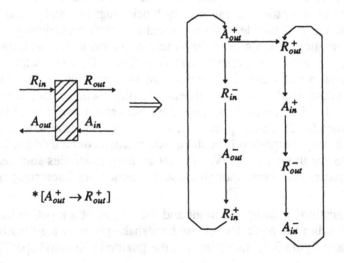

**Figure 3-8.** Another pipeline interconnection circuit specified by $*[A_{out}^+ \to R_{out}^+]$ and its graphical representation.

$$*[A_{out}^+ \to R_{out}^+], \tag{3.2}$$

where the pre-condition is replaced by the acknowledge signal $A_{out}$. It will be shown in Section 3.5 that the circuit in figure 3-7 results in only a 50%

hardware utilization of computation blocks in a pipelined architecture, because only alternate blocks can compute concurrently, while the circuit specified in figure 3-8 allows every block to compute concurrently. Our synthesis technique therefore leads immediately to an improvement in the performance of a self-timed pipeline. The analyses of these two specifications will be given in the next section along with the reasons for the performance discrepancy.

## 3.4. WEAKEST SEMI-MODULAR CONSTRAINTS

Guarded commands specify the necessary conditions for the behavior of a circuit that implements an interconnection block. To bridge the gap between this behavioral specification and its circuit implementation, we need only strengthen these pre-conditions until they are also sufficient to form a *semi–modular* circuit, as defined below (a modified definition from [3]):

> **Definition 3.4.** A signal transition is semi-modular if once the transition is enabled, only the firing of the transition can disable it.

By disabling a transition we mean that the enabling condition no longer holds. For example, if $x^+$ enables $y^+$, semi-modularity requires that only after $y$ has actually gone high ($y^+$), can $x$ go low ($x^-$). A circuit is semi-modular if every signal transition in the circuit is semi-modular. We picked this definition as the objective for synthesis for many reasons. First, semi-modularity is defined on signal transitions, which are much fewer than states in number. In fact, the algorithm we have developed is capable of deciding the existence of a hazard-free implementation, given a behavioral specification, within polynomial steps. Second, semi-modularity is in general stronger than speed-independence, since semi-modularity implies a live strongly-connected graph. Third, it bears strong correlation with the hazard definition and, as we will show in the next few paragraphs, there exists an equivalence between hazard-free circuits and semi-modular signal transition graphs under certain broad conditions, which are always satisfied in the synthesis procedure by construction.

Semi-modularity is sometimes confused with the *persistence* condition [8,2] in the Petri net analysis. Since a signal transition graph is an interpreted Petri net, the persistence property defined on uninterpreted Petri nets cannot describe the *disabling* condition. Hence semi-modularity is a stronger condition than the property of persistence.

A totally sequential circuit is semi-modular by definition because only one signal transition is enabled at a time. Although this restriction simplifies the behavior of a circuit, it may often be desirable to allow more than one signal transition to be simultaneously enabled, thus allowing a type of concurrency in the circuit. But concurrency also introduces potential circuit hazards. The degree of concurrency under the hazard-free condition is a measure of optimality in our self-timed circuit synthesis.

## 3.4.1. Consistent State Assignments

We now describe the relationship between semi-modularity and the circuit hazard condition. The liveness condition in a signal transition graph insures that every live-safe marking has a bitvector state assignment, but this bitvector might not be consistent with a given token marking. A token on an arc means that the enabling signal transition has taken place and the signal level in the bitvector (the state) is thus determined. A bitvector state assignment is said to be consistent with a live-safe marking if each signal level can be uniquely defined by token positions. For example, the marking on the live signal transition graph shown in figure 3-5(a) is live and safe, and has a bitvector state assignment $(0, 0, 1)$. Signal $x$ was assigned low in this state since transition $x^-$ was fired after transition $x^+$, according to the signal transition graph; however, this assignment is not consistent with the token on $x^+ \rightarrow y^+$, which states that $x$ should be high. Semi-modularity links a live-safe marking with a consistent state assignment.

> **Claim 3.6.** In a live signal transition graph, every live-safe marking has a consistent state assignment if and only if the signal transition graph is semi-modular.

To show the sufficient condition, assume that every life-safe marking in a signal transition graph has a consistent state assignment. If a transition $t$ is enabled, then all the incoming arcs to $t$ are marked, which assign each enabling signal a definite level. Since every live-safe marking assigns a consistent level to each signal, the enabling signals will not be assigned different levels before $t$ is fired; in other words, the enabling condition will not be disabled before $t$ is fired and therefore $t$ is semi-modular. The necessary condition can be proved by contradiction.

Since circuit behavior is described by signal transitions and states are specified by token markings on a signal transition graph, every marking corresponding to a consistent bitvector assignment means that circuit behavior (transition sequences) can be uniquely defined on signal levels (states in state diagrams). Consequently the circuit can be implemented with level-triggered logic. Semi-modularity insures that for each enabled transition, the enabling signal levels must hold before the transition actually fires. This is the first step toward a hazard-free level-triggered logic implementation.

Before we proceed with the discussion of hazard-free circuits, a very important assumption, the environmental constraints, as imposed by the complete circuit model needs to be satisfied. For each signal in a complete circuit, there is a partial circuit characterized by a Boolean function with the signal as the sole output, as shown in figure 3-6. The reason we insist on a complete circuit representation (using repetitive constructs) is to insure proper control over the input signals to the partial circuits being synthesized; otherwise a hazard-free implementation wouldn't be possible. An interconnection circuit implements only part of the complete circuit; other parts of the circuit are performed by external networks, which implement the environmental constraints with unknow delays.

From our definition of a hazard the output signal of a hazard-free circuit should behave as predicted by the state diagram for all the allowed input transition patterns. Now we show the relation between this hazard definition and consistent state assignments.

Although we often divide various circuit hazards into different categories, the origin of a hazard is always the same: when more than one signal transition, internally or externally, are enabled, these signal transitions disagree on the output signal level. A speed-independent circuit as defined by Definition 3.3 allows simultaneous signal transitions, and specifies that every enabled transition will eventually be fired. However, this definition only *describe* the circuit behavior and does not tell how to *implement* such a circuit.

Shown in figure 3-6 is the state diagram and Boolean functions derived from the speed-independent circuit in figure 3-5. The Boolean function for signal $y$ is $x + z$. If the sequence of transitions $x^+ \rightarrow z^+ \rightarrow x^-$ take place before $y$ goes high, the $x$ input to the OR gate of $y$ would go low before the enabled transition $y^+$ is actually fired. The enabling condition $(x^+)$ for $y^+$ is disabled and the enabling responsibility is transferred to $z^+$. If $y$ is about to be set high by $x^+$ when $z^+ \rightarrow x^-$ happens, the OR gate may display a spiky output and the circuit behavior deviates from that predicted by the state diagram.

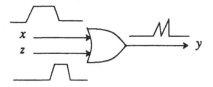

**Figure 3-9.** An OR-gate hazard caused by transferring enabling transitions.

This kind of hazard is described in [39], which happens in real circuits if the overlap time of $x^+$ and $z^+$ is small, as shown in figure 3-9. The reason is that $x^-$ and $z^+$ tend to drive the output $y$ towards different levels. The physical manifestation is that state $(0,0,1)$ is not consistent with its live-safe marking in the signal transition graph, the marking shown in figure 3-5(a).

The signal transition graph shown in figure 3-5 is not semi-modular. Transition $y^+$ is enabled by $x^+$, but the enabling condition $x^+$ can be disabled ($x$ going low) before $y$ actually goes high. If we add an arc directed from $y^+$ to $x^-$ to prevent $x$ from going low before $y$ goes high, as shown in figure 3-10(a), then the enabled transition $y^+$ cannot be disabled ($x$ going low) until $y$ has actually gone high. The newly-added arc is the weakest condition (necessary and sufficient) that imposes semi-modularity on transition $y^+$. The necessary and sufficient condition can be proved by contradiction. Other arcs, such as the one directed from $y^+$ to $z^+$, also suffice to make the transition semi-modular, but are too strong in that they prevent the possible concurrent operation of $y^+$ and $z^+$.

The state diagram shown in figure 3-10(b) is derived from the semi-modular signal transition graph shown in figure 3-10(a), where every live-safe marking corresponds to a consistent state assignment. State $(0,0,1)$ is deleted from the state diagram shown in figure 3-10(b) since the semi-modular

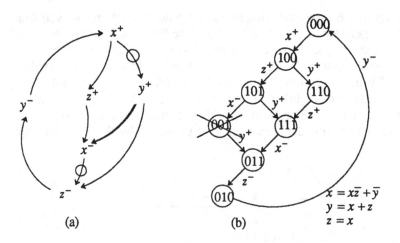

(a)                      (b)

$$x = x\bar{z} + \bar{y}$$
$$y = x + z$$
$$z = x$$

**Figure 3-10.** (a). The semi-modular signal transition graph derived from figure 3-5(a) in which the arc directed from $y^+$ to $x^-$ (the double-lined arc) is the weakest semi-modular constraint to make transition $y^+$ semi-modular. (b). The state diagram derived from figure 3-10(a). Semi-modularity insures that every state in the state diagram is consistent with its corresponding token marking.

condition $y^+ \rightarrow x^-$ makes the corresponding marking shown in figure 3-10(a) unsafe. The Boolean functions for the three signals $x$, $y$, and $z$ are also shown in figure 3-10(b).

As level-triggered logic is not capable of *remembering* that a transition had happened when the signal level indicates otherwise, consistent state assignments are necessary to insure that signal levels are not to be changed to confuse the gate which controls the enabled transition. If all signal transitions agree on driving each signal towards the same level at all time, circuit hazards will not happen. It can be checked that all allowed concurrent transitions in the Boolean functions of figure 3-10(b) drive each output signal toward the same level.

## 3.4.2. Maximum Concurrency

Since every added arc in a signal transition graph represents a possibility of forming more simple loops, imposing semi-modularity on a circuit reduces the number of legal life-safe markings in a signal transition graph. Arcs with weakest semi-modular conditions eliminate only those live-safe markings which do not have a consistent state assignment. For the circuit shown in figure 3-10(a), a marking of tokens on $x^+ \rightarrow y^+$ and $z^+ \rightarrow x^-$ is a live-safe marking if the weakest semi-modular condition $y^+ \rightarrow x^-$ is added to the graph. But a stronger semi-modular condition $y^+ \rightarrow z^+$ would announce this marking unsafe, since the simple loop $x^+ \rightarrow y^+ \rightarrow z^+ \rightarrow x^- \rightarrow z^- \rightarrow y^- \rightarrow x^+$ would contain two tokens.

In Section 3.5, we will show that a signal transition graph with the weakest semi-modular constraints can be constructed by a polynomial time algorithm (polynomial in the number of signals in the circuit), and thus the existence of a hazard-free implementation given a circuit behavioral specification can be determined in polynomial time. Consequently, we do not need combinatorial algorithms, which have exponential complexity in the number of signals in the circuit (for example the use of state diagrams), until the existence of a hazard-free implementation is assured.

The degree of concurrency in a complete circuit is related to the number of states. The weakest semi-modular constraints allow the maximum number of live-safe markings that can be defined on a semi-modular graph derived from the same repetitive guarded command, and the corresponding state diagram contains the largest set of consistent states and the greatest concurrency.

To derive the weakest semi-modular conditions in a live signal transition graph, non-semi-modular transitions must first be identified. Transitions that cause potential non-semi-modularity are those that enable multiple allowed sequences. The weakest semi-modular conditions of a non-semi-modular transition is the arcs directed from the non-semi-modular transition

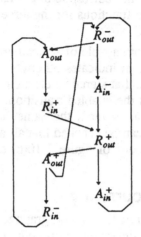

**Figure 3-11.** Constructing the semi-modular transition graph of the interconnection circuit specified by $*[R_{in}^+ \rightarrow R_{out}^+]$ with weakest conditions.

to all the transitions whose inverse transitions enable this non-semi-modular transition. After the weakest conditions for a particular non-semi-modular transition are added to the original graph, redundant arcs can be deleted. Semi-modularity should be checked recursively after newly-added arcs until every transition in a signal transition graph is semi-modular.

For illustration, take the circuit in figure 3-7, which is the signal transition graph derived from the guarded command $*[R_{in}^+ \rightarrow R_{out}^+]$. There is only one transition, $R_{in}^+$, which has multiple allowed sequences. Therefore an arc directed from $R_{out}^+$ to $R_{in}^-$ should be added to the graph to make transition $R_{out}^+$ semi-modular. But $R_{in}$ is an input signal. A physical circuit cannot impose constraints on its input signals; it can only impose constraints on the output signal that in turn influence the target input signal. Hence the first arc added to the graph is directed from $R_{out}^+$ to $A_{out}^+$ as shown in figure 3-11, after which the redundant arc from $R_{in}^+$ to $A_{out}^+$ is deleted. There is then a new non-semi-modular transition, $A_{out}^+$. Therefore an arc directed from $A_{out}^+$ to $R_{out}^-$ is added. The newly added arc in turn causes $R_{out}^-$ to be non-semi-modular. Therefore another arc directed from $R_{out}^-$ to $A_{out}^-$ is added, making the circuit semi-modular.

## 3.5. SYNTHESIS PROCEDURE AND THE C-ELEMENT

A deterministic algorithm to synthesize self-timed hazard-free interconnection circuits will now be described.

### 3.5.1. The Synthesis Procedure

The input to the synthesis program is a guarded command and the output of the program is a Boolean function for each output handshake signal. The guarded command is first translated to a complete circuit by adding transitions to enforce the four-phase handshake protocol. Since the translated graph is live (enforced by the handshake protocol), we need only consider the property of semi-modularity. Arcs of weakest semi-modular conditions are added to the circuit, from which a state diagram with consistent state assignments is formed. Standard Boolean minimization tool [40] is used to give a Boolean expression for each output signal. A circuit implementation will then be derived from these Boolean expressions.

To illustrate the synthesis process, the signal transition graph in figure 3-11 is translated to a state diagram shown in figure 3-12(a). A state is defined as $(R_{in}, R_{out}, A_{in}, A_{out})$. There are four partial circuits to be synthesized, but the partial circuits corresponding to $A_{in}$ and $R_{in}$ need not be considered since they are inputs to the interconnection circuit. We are only interested in the partial circuits corresponding to $R_{out}$ and $A_{out}$, which are the outputs of the interconnection circuit. From the state diagram in figure 3-12(a), we can construct a K-map for each of the output signals $R_{out}$ and $A_{out}$. The element corresponding to a location (a state) in a K-map is the level of the output signal in the next state. If there exist multiple allowed next states and these next states assign both levels to the output signal, the level that is different from the present output level is put in the location. The K-maps for signals $R_{out}$ and $A_{out}$ are drawn in figure 3-12(b) along with their Boolean functions. A logic realization is shown in figure 3-13. We will discuss the hazard properties of the $SR$-latch after we introduce the C-element.

The efficiency of hardware utilization can be derived from the signal transition graph. In a live semi-modular graph, once a request is accepted, each transition must be traversed once and only once before a second request can be accepted. The circuit delay in response to an external request is the longest delay of all the simple loops in the graph. Since computational delays are by assumption much longer than handshake signal transitions, we first identify computation arcs (arcs corresponding to latencies of computation blocks) to be the arcs directed from $R_{out}^{+}$ to $A_{in}^{+}$ and from $A_{out}^{-}$ to $R_{in}^{+}$. Redrawing the transition graph in figure 3-11 with computation arcs represented by squiggled lines, as shown in figure 3-14(a), we can see that there is a loop of transitions $A_{out}^{-} \rightarrow R_{in}^{+} \rightarrow R_{out}^{+} \rightarrow A_{in}^{+} \rightarrow R_{out}^{-} \rightarrow A_{out}^{-}$

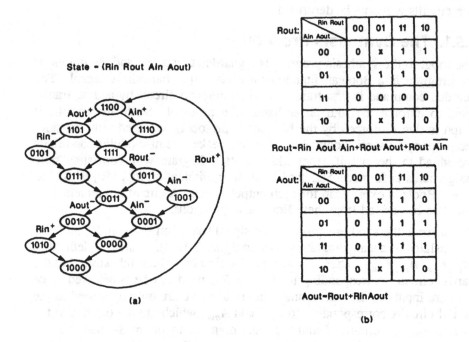

Figure 3-12. (a). The circuit state diagram derived from the semi-modular signal transition graph shown in figure 3-11. (b). The K-maps of output signals $R_{out}$ and $A_{out}$ derived from the state diagram in figure 3-12(a).

which traverses two computation arcs. If we assume that every computation block takes the same delay to complete its task, the handshake circuit specified by $*[R_{in}^+ \rightarrow R_{out}^+]$ will allow only half of computation blocks process data at the same time. We call this handshake scheme a *half-handshake*.

The semi-modular transition graph specified by $*[A_{out}^+ \rightarrow R_{out}^+]$ is shown in figure 3-14(b). There is no loop that contains both computation arcs; hence this circuit allows every computation block to process data at the same time. We call this handshake scheme a *full-handshake*. As shown in figure 3-14(a), the reason that the circuit specified by $*[R_{in}^+ \rightarrow R_{out}^+]$ results in a half-handshake is that the arc $R_{out}^- \rightarrow A_{out}^-$, added by semi-modular constraints to avoid a hazard condition, forces sequential computation of the

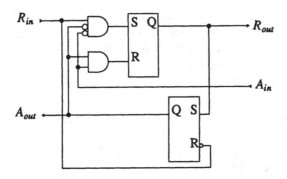

**Figure 3-13.** Logic implementation of the handshake circuit specified by $*[R_{in}^+ \rightarrow R_{out}^+]$. Of all the circuits that behave correctly according to the specification, this circuit allows maximum concurrent operation.

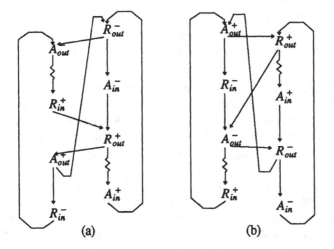

(a)                                    (b)

**Figure 3-14.** (a). The transition graph in figure 3-11 with computation arcs represented by squiggled lines. There is a loop of transitions $A_{out}^- \rightarrow R_{in}^+ \rightarrow R_{out}^+ \rightarrow A_{in}^+ \rightarrow R_{out}^- \rightarrow A_{out}^-$ which traverse both computation arcs. Therefore this handshake circuit results in at most 50% hardware utilization of computation blocks in a pipelined architecture. (b). The semi-modular transition graph derived from the handshake circuit specified by $*[A_{out}^+ \rightarrow R_{out}^+]$. There is no loop in this graph that contains both computation arcs. Therefore this handshake circuit allows every computation block to be computing at all time.

two blocks. The semi-modular constraints added to figure 3-14(b) do not constitute such a constraint.

In fact the behavior specification of $*[A_{out}^+ \rightarrow R_{out}^+]$ does suggest a full-handshake operation. As the pre-condition is the acknowledge signal $A_{out}$, an indication to block A that block B has received the output datum from block A, block A can proceed with processing the next datum while block B is processing the present one; consequently both blocks can process data at the same time.

The difference between the half- and full-handshake demonstrates one of the advantages of a high-level specification for circuit behavior, since slight differences in specifications may result in very different circuit performance. A pipeline interconnection circuit has only four handshake signals. Imagine that for an interconnection circuit with 10 signals (the simplest multiplexer as shown in the next section has nine signals), there will be at most $10 \times 9 = 90$ arcs in its transition graph and $2^{90}$ possible specifications to choose from in a transition specification. Other implementations of four-phase handshake circuits [37,32,7] are examples of different specifications from the same behavioral description. A simple high-level description such as guarded commands reduces the decision space tremendously and thus allows us to exercise reasoning and analysis at the circuit behavioral level, rather than directly specifying signal transitions when the corresponding circuit operation is difficult to foresee.

The logic implementation of the full-handshake is shown in figure 3-15. Inspecting the Boolean expressions for the full-handshake, the circuit is actually formed by two C-elements. A two-input C-element implements the Boolean function $C = AB + BC' + AC'$, where $A$ and $B$ are the input signals, $C'$ is the previous output signal, and $C$ is the current output signal [3]. The C-element has the property that the output signal will change to the input level when both inputs are of the same level; otherwise the output stays unchanged. The C-element is one of the basic units in the early approaches to designing speed-independent circuits. We did not start with the assumption of using logic modules, not even C-elements, but two C-elements arise naturally from the synthesis procedure. Since two C-elements constitute a full-handshake, one might think that one C-element would be a half-handshake. One C-element with an inverter is a correct half-handshake (alternate computation blocks processing at the same time), but it does not allow the same amount of concurrency as displayed by the circuit shown in figure 3-13, in which concurrent reset of the two computation blocks is possible.

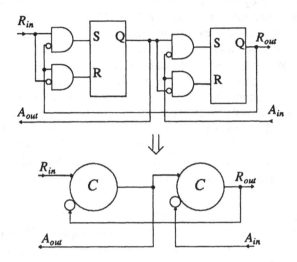

**Figure 3-15.** Logic implementation of the full-handshake specified by $*[A_{out}^{+} \rightarrow R_{out}^{+}]$. This figure shows that two C-elements connected in such a way is the design for a full-handshake circuit.

## 3.5.2. SR-Latches vs. C-Elements

The logic functions of the half-handshake shown in figure 3-13 uses two $SR$-latches. In classic logic design, if the Boolean function of an $SR$-latch is written as $Q = S + RQ'$, where $S$ and $R$ are some combinational functions of input signals and $Q'$ is the previous output signal, then $S$ and $R$ cannot both be high at the same time, since the condition $S = R = 1$ would set the output $Q$ undefined (a metastable condition) and represent a circuit hazard. This problem is usually circumvented by designing set-dominant (or reset-dominant) $SR$-latches in which the outout $Q$ is set high (or reset low) whenever $S$ (or $R$) is high. However, because of the assumed unbounded gate delays, there are situations in which the values of $S$ and $R$ cannot be predicted by the Boolean function. Hence the choice of a set-dominant $SR$-latch or a reset-dominant $SR$-latch becomes a function of gate delays.

Since logic gate delays are assumed to be finite but unbounded in speed-independent circuits, $S$ and $R$ may both become high with different gate delay assumptions. Take the $SR$-latch in figure 3-13 that controls $R_{out}$ for example: if the delay of the AND gate at the input of $S$ is longer than the delay of the AND gate at the input of $R$, both $S$ and $R$ as seen by the $SR$-latch can be high at the same time, which represents a mismatch between the circuit behavioral description and its logic implementation [41].

If we chose to abide by the orthodoxy of a pure speed-independent design, any logic implementation incorporating $SR$-latches cannot be truly speed-independent, as the mutual exclusion of $S$ and $R$ cannot be guaranteed through unbounded gate delays. C-elements do not have this problem and it has been proposed that speed-independent (and delay-insensitive) circuits use only C-elements as memory elements [42]. Since the $SR$-latches in the Boolean functions derived from a semi-modular circuit preclude any hazard condition by allowing only those input transitions which agree on the output level to be enabled at the same time, the functionality of an $SR$-latch is equivalent to that of a C-element with an inverter. Therefore C-elements can be used to replace $SR$-latches without any functional difference. The resulting circuit for the half-handshake, shown in figure 3-16. is hazard-free under all variations of gate delays.

The above discussion of hazard-free circuits depends on the existence of an *atomic* C-element, whose output follows the two inputs with a finite delay regardless of the temporal relations between the two inputs. An atomic C-element cannot be implemented by connecting basic logic gates, as the implementation will then contain potential hazards such as contained in an $SR$-latch. Several circuits have been proposed for implementing the atomic C-element at the transistor level [43,16]. A six-transistor dynamic C-element used in our design is shown in figure 3-17. The ideal behavior of an atomic C-element can be attained if transmission delays *within* the element are negligible compared to gate delays, as is usually, if not always, satisfied in integrated circuit design.

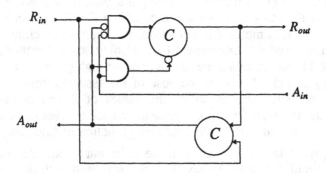

**Figure 3-16.** A hazard-free implementation of the half-handshake interconnection circuit.

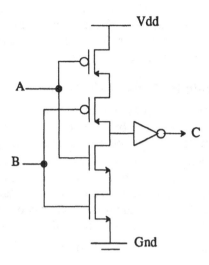

**Figure 3-17.** An atomic C-element implemented by six transistors.

### 3.5.3.  The Synthesis of OR Constructs

Since the weakest conditions for semi-modularity have the basic assumption of using AND constructs, OR constructs and alternative constructs exhibit a special problem in the synthesis algorithm. Since OR constructs can be translated to equivalent alternative constructs, alternative constructs are used internally in our synthesis procedure. An example of the OR construct is a multiplexer circuit (see Section 6). An alternative construct allows one of several alternative paths to be taken depending upon external conditional signals. This construct defines several different networks, which can change as a function of time, depending on the state of the conditional signals. The alternative construct is thus a means for compactly specifying these various networks. In our synthesis procedure we check semi-modularity for each of the possible networks in parallel and use an AND operation with the conditional signals to specify the actual path and an OR operation to combine the various paths together (a free-choice Petri net). State diagrams are then constructed with conditional signals specifying possible networks, on which the final logic synthesis is based.

### 3.5.4.  The Automated Synthesis Program

An automated self-timed circuit synthesizer has been implemented in C and

Franz Lisp under Unix[3] 4.2bsd. The synthesis program consists of three parts: constructing a semi-modular signal transition graph from a guarded command, translating a signal transition graph to a state diagram, and generating a Boolean expression for every output signal from the state diagram.

## Guarded Commands to Semi-Modular Signal Transition Graphs

The input to the synthesis program is a guarded command set. Each guarded command is represented by a list of partial orderings specified by the pre-conditions and the statement list. For example, $*[A_{out}^+ \rightarrow R_{out}^+]$ is represented as $(A_{out}^+ \ R_{out}^+)$; $A_{out}^+$ is called the enabling transition and $R_{out}^+$ the enabled transition. OR constructs are first translated to alternative constructs and each clause in an alternative construct is handled independently in this phase; the fact that the clauses in an alternative construct are related by conditional signals is taken into account at the state diagram level.

The procedure for translating a guarded command to a semi-modular signal transition graph is:

**Step 1.** For every interconnection signal specified in the guarded command, sequences of transitions implicitly assumed by using the four-phase handshake protocol are included in a list called *transition_graph*. Conditional signals, which are data signals from a controller rather than handshake signals, are identified at this point and the list *transition_graph* actually contains multiple sub-graphs if there are conditional signals specified.

**Step 2.** The partial orderings represented by the guarded command are added into the *transition_graph* list. For every partial ordering added into the *transition_graph*, the enabling transition is put into a list called *non_semi_modular_nodes*, containing transitions which enable multiple allowed sequences. Each subgraph in *transition_graph* has its own *non_semi_modular_nodes* list since semi-modularity will be checked independently for each subgraph.

**Step 3.** For each enabling transition in the list of *non_semi_modular_nodes*, semi-modularity is checked by tracing the graph from each enabled transition. If there is a loop, starting from an enabled transition, which contains both the enabling transition and its inverse transition, then the enabled transition is semi-modular. A recursive *trace* function can easily detect the existence of such a loop.

---

[3] Unix is a trademark of AT&T and bsd stands for Berkeley software distribution.

If there are no such loops, an arc directed from the enabled transition to the inverse transition of the enabling transition is added to the *transition_graph* list. The enabled transition is appended to the *non_semi_modular_nodes* as it enables multiple sequences with the added arc. Signal types (input or output) are checked at this point to insure that added arcs are always directed to output signals. The procedure is recursively called until the *non_semi_modular_nodes* list becomes empty.

The computational complexity of Step 1 and Step 2 is low. The complexity of Step 3, which imposes weakest semi-modular constraints onto a signal transition graph, is at most $N^2$ times the complexity of the *trace* function, where $N$ is the number of signals involved. The complexity of the *trace* function is bounded by $O(N^2)$ using a depth-first search algorithm. The result is a polynomial time algorithm.

## Signal Transition Graphs to State Diagrams

Since the state diagram of a live-safe circuit must be strongly connected, there are neither initial states nor final states. We need only construct the partial orderings which represent reachable states. A state of a semi-modular circuit can be represented by a live-safe marking in a signal transition graph. For example, the state $(R_{in} R_{out} A_{in} A_{out}) = (0011)$ in figure 3-12(a) can be represented by a set of tokens marked on the arcs of $R_{in}^- \rightarrow A_{out}^-$, $R_{out}^- \rightarrow A_{out}^-$, and $R_{out}^- \rightarrow A_{in}^-$ in the corresponding transition graph shown in figure 3-11. The allowed signal transitions from the present state are the transitions with all of their input arcs marked. The next states can be constructed by allowing the marking tokens to free-flow, one transition as a time. Any state that can be reached within one transition of the present state is a next state. A don't-care or unreachable state does not have a corresponding live-safe marking; its corresponding marking will eventually cause deadlock.

The algorithm can be summarized as follows:

**Step 1.** Pick a starting state, say all signals low, and translate it to a marking, a set of tokens, on the signal transition graph. A live-safe marking can be checked by the property that each arc is in a simple cycle with exactly one token [4]. If the marking is live and safe, append it to an empty list called *present_state*. Copy *present_state* to *reachable_states* and go to Step 2. If the marking is not live and safe, randomly pick another state and repeat Step 1.

**Step 2.** For every marking in *present_state*, the possible next state markings can be generated by letting the present state marking free-flow. The present state marking and the possible next state markings form a list of partial orderings and are stored in the *state_diagram* list. For all those possible next state markings which were not already in the *reachable_states* list, append them to both the *reachable_states* list and the *present_state* list. Delete the present state marking from the *present_state* list. Repeat Step 2 until the *present_state* list is empty.

**Step 3.** Subgraphs of alternative constructs are combined at this phase. Translate markings in *reachable_states* and *state_diagram* to state representations. Condition signals are added into the bitvector in *state_diagram* and *reachable_states* so that each guarded command in the command set corresponds to exactly one state diagram. Finally, construct a *don't_care_states* list which contains all the other states not in the *reachable_states* list.

The complexity of Step 1 can usually be ignored. The algorithm used in Step 2 is of complexity $O(2^N NE)$ in finding all possible next states, where $N$ is the number of signals and $E$ is the number of arcs in a transition graph ($E = N (N-1)$ at most). To check if a possible next state is in *reachable_states*, a binary search algorithm is used and therefore the complexity of constructing the *reachable_states* list is of $O(2^N N)$. The complexity of Step 3 is $O(2^N)$.

## State Diagrams to Boolean Expressions

Standard Boolean minimization is used in this phase of the synthesis. Each partial ordering in the *state_diagram* list represents the present state and possible next states. Since states are composed of input and output signals, output signal levels can be read out from states.

**Step 1.** For each output signal, we need to construct a current-to-next-state mapping. At each state, if there exist multiple next states and these states assign both levels to the output signal, the level that is different from the present output level is given as the next signal level. The states with the next signal level equal to 1 are selected to form a list called *one_states*.

**Step 2.** Each element in the *one_states* list is translated to a Boolean expression and so does each element in the *don't_care_states* list.

**Step 3.** Use a Boolean minimization tool such as ESPRESSO [40] to translate the Boolean function of each output signal to its minimal representation.

The computational complexity of Step 1 and Step 2 is on the order of $O(2^N N)$, since there are at most $2^N$ states. Step 3 is a standard combinatorial minimization problem and constitutes the speed bottleneck of the synthesis process, if *exact* minimization is enforced. Contraction was proposed for more efficient Boolean minimization [44], but would adversely introduce heuristics and produce suboptimum solutions.

Since we need only to synthesize each guarded command once and the synthesized logic can be reused as a macrocell, usually we chose to exercise exact Boolean minimization for a minimum design.

## 3.6. INTERCONNECTION DESIGN EXAMPLES

The synthesis program has been used to design various digital architectures, including a general purpose programmable processor and bus interface circuits. In this section, we will show some examples of how to specify interconnection circuits with guarded commands, and present the resulting synthesized circuits. More complicated synchronization circuits used for memory interface and bus control can be found in the next chapter. Since the computational complexity of imposing semi-modularity is polynomial in the number of signals, the synthesis run-time performance is dominated by the Boolean minimization tool used (e.g. ESPRESSO).

In the following examples, the implicit loop of the four-phase handshake is assumed in all circuits. If not explained explicitly, signals shown in the figures of this section are all handshake signals and registers are ignored for simplicity.

### 3.6.1. Pipeline and Non-Pipeline Connections

Pipelining increases system throughput by introducing sample delays in the datapath. A sample delay can be implemented by inserting a register in a full-handshake circuit, as will be discussed in Chapter 4. However, sometimes it is necessary to connect two stages of computation blocks without an intermediate sample delay. A simple example is shown in figure 3-18 where block A and block B are operated sequentially. The interconnection circuit between these two computation blocks can be specified by a sequential construct

$$*[ R_{in}^+ \rightarrow R_{out}^+ ; A_{in}^+ \rightarrow A_{out}^+ ] \qquad (3.3)$$

to indicate the fact that only after block B has completed its task can block A accept a new datum. The logic implementation of the specification is shown in figure 3-18(b). The interconnection circuit shown in figure 3-18(b) is not simply a half-handshake circuit, because alternate blocks would then be computing concurrently, as opposed to a non-pipelined interconnection of blocks, which only allows one block to be active at a time.

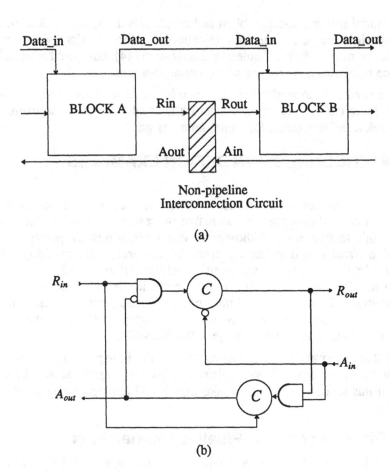

**Figure 3-18.** (a). A non-pipeline interconnection circuit between two computation blocks A and B. A sequential construct $*[R_{in}^+ \rightarrow R_{out}^+ ; A_{in}^+ \rightarrow A_{out}^+]$ is used to indicate that only after block B has completed its task can block A accept a new datum. (b). Logic implementation of the interconnection circuit specified in (3.3).

As will be discussed in the next chapter, it can be verified that the non-pipeline interconnection circuit does not need a pipeline register between the two computation blocks. A simplier circuit to connect two sequential computation blocks would be to connect $R_{in}$ to $R_{out}$ and $A_{in}$ to $A_{out}$. However, the circuit in figure 3-18(b) allows a form of concurrency in that block A can start computing the next datum while block B is being reset. The simple connection allows only sequential operation and thus is not the best design.

## 3.6.2. Multiple-Input Block

Computation blocks with two input sources can be modeled as shown in figure 3-19(a), where the guarded command specification for the interconnection circuit with a pipeline stage is

$$*[A_{out1}^+ \cap A_{out2}^+ \rightarrow R_{out}^+]. \tag{3.4}$$

$A_{out1}^+$ and $A_{out2}^+$ are ANDed in the pre-conditions to specify that only after both of the input data are valid can the next block start computation ($R_{out}^+$). In this specification, pipeline registers are needed between the two stages of computation. The logic implementation of the specification is shown in figure 3-19(b).

To transfer multiple-input data without forming a pipeline stage, a sequential guarded command can be used:

$$*[R_{in1}^+ \cap R_{in2}^+ \rightarrow R_{out}^+ ; A_{in}^+ \rightarrow A_{out1}^+ , A_{out2}^+]. \tag{3.5}$$

The logic implementation is shown in figure 3-19(c). Blocks with more than two inputs can be synthesized in a similar way.

## 3.6.3. Multiple-Output Block

Computation blocks with two output destinations can be modeled as shown in figure 3-20(a), where the guarded command specification for the connection circuit with a pipeline stage is

$$*[A_{out}^+ \rightarrow R_{out1}^+ , R_{out2}^+]. \tag{3.6}$$

In this specification, the statement list contains two actions, $R_{out1}^+$ and $R_{out2}^+$, which are to be performed concurrently. By this specification a pipeline register is necessary between the two stages of computation. The logic implementation synthesized for this specification is shown in figure 3-20(b).

If a pipeline register is not desired, a sequential construct can be used.

## 3.6.4. Multiplexer

A two-input multiplexer is shown in figure 3-21(a) with the guarded command specification of

$$*[(A_{out1}^+ \cap C_{in}^+ \cap T) \cup (A_{out2}^+ \cap C_{in}^+ \cap \bar{T}) \rightarrow R_{out}^+], \tag{3.7}$$

where $C_{in}$ is the request signal from the controller, $C_{out}$ the acknowledge signal fed back to the controller, and $T$ the conditional signal (guarded by $C_{in}$) to select one of the input blocks. $C_{in}^+, A_{out1}^+$, and $A_{out2}^+$ are specified in the pre-condition to insure that no operations can be performed until the conditional signal $T$ and the selected datum are valid at the multiplexer input and that the full-handshake configuration is used. The guarded command in (3.7) specifies that if $T$ is high, the datum from block A is to be

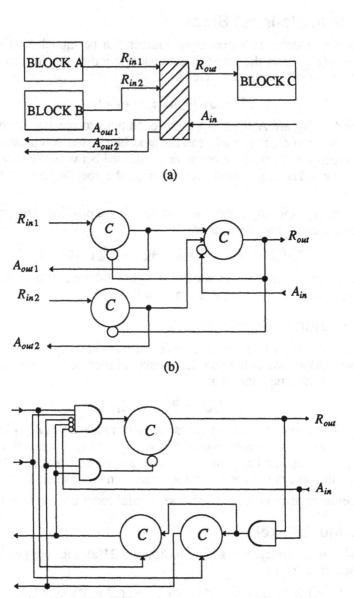

**Figure 3-19.** (a). A computation block with two input sources. (b). Logic implementation of the interconnection circuit with pipeline registers. (c). Logic implementation of the interconnection circuit without explicit pipeline registers.

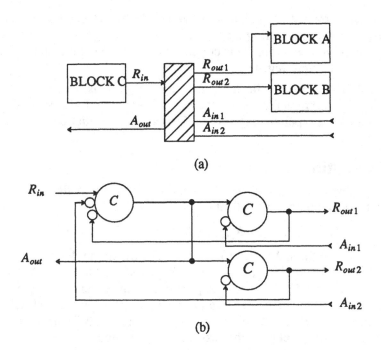

**Figure 3-20.** (a). A computation block with two output destinations. (b). Logic implementation of the interconnection circuit with pipeline registers.

transferred to block C and that if $T$ is low, the datum from block B is to be transferred. If both of the input blocks hold valid data in the same instruction, the unselected datum will be sitting at the multiplexer input waiting to be selected in a later instruction. In a self-timed processor, multiplexers and demultiplexers must reflect the architectural decision on data flow control by properly specified guarded commands. Another example of multiplexers in which the unselected data would be overwritten will be given in the next chapter.

Only the interconnection part of a multiplexers function is shown in figure 3-21(a). The actual multiplexing operation can be performed by a DCVSL block controlled by $R_{out}$ and $T$.

An OR construct is used in the multiplexer specification. The OR construct is first translated to an alternative construct

$$* [A_{out1}^+ \cap C_{in}^+ \cap T \to R_{out}^+ \mid A_{out2}^+ \cap C_{in}^+ \cap \bar{T} \to R_{out}^+ ]. \qquad (3.8)$$

Semi-modularity is checked individually on the two transition graphs and a state diagram is constructed by combining the two. The synthesized multiplexer interconnection is shown in figure 3-21(b). The conditional signal $T$ must remain stable during the transitions $C_{out}^+ \rightarrow C_{in}^- \rightarrow C_{out}^-$ to ensure a hazard-free implementation. This condition on $T$ can be guaranteed if the controller uses an output register.

With the specification given in figure 3-21(a), a register is needed between the two stages of computation since the multiplexer circuit also serves as a pipeline stage. If a pipeline stage is not desired, a sequential construct can be used.

## 3.6.5. Demultiplexer

A demultiplexer with two output destinations is shown in figure 3-22(a) with the guarded command specification of

$$*[A_{out}^+ \cap C_{in}^+ \cap T \rightarrow R_{out\,1}^+ \,|\, A_{out}^+ \cap C_{in}^+ \cap \overline{T} \rightarrow R_{out\,2}^+ ], \qquad (3.9)$$

where $T$ is the conditional signal to select one of the output blocks. In this specification, the datum from block C is to be transferred to block A if $T$ is high and to block B if $T$ is low. $A_{out}^+$ and $C_{in}^+$ are specified in the preconditions for the same reason as in a multiplexer. The synthesized demultiplexer interconnection circuit is shown in figure 3-22(b). In this specification a register between the two stages is assumed.

In this chapter we have shown that automatic synthesis of self-timed interconnection circuits from a high-level specification is feasible through the separation of computation and interconnection modules. By providing a modular design framework in which designers can prototype their systems based entirely on local properties without worrying about global variations, we believe that the effort required in implementing complex digital systems can be greatly reduced.

Designing efficient combinational circuits with completion signal generation is a challenging task for circuit designers, but potentially rewarding since systems built on such circuits will achieve processing speeds closer to the potential of future technologies. Without global synchronization, block diagrams can be converted into working systems with less difficulty, shortening the time between system design and physical implementation.

Guarded commands provide a high-level interface for specifying interconnection circuits; however, when systems become complex, even specifying interconnection using guarded commands could be a painstaking task. It would be interesting to extend the synthesis procedure to accommodate architectural information so that proper guarded commands can be systematically specified. We thus suggest future research into hardware description languages. Given the system architecture and instruction operations, guarded commands for each interconnection block should be generated

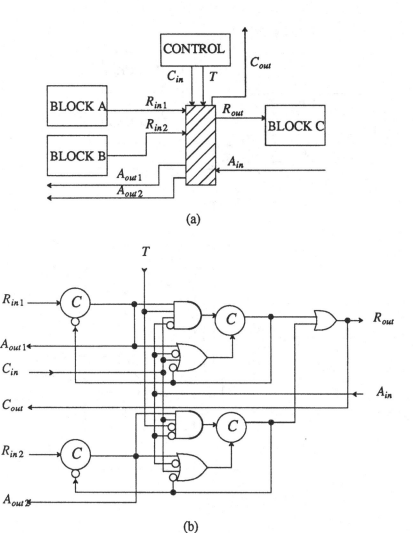

(a)

(b)

**Figure 3-21.** (a). A multiplexer interconnection circuit with two input sources. (b). Logic implementation of the multiplexer interconnection circuit specified in (3.7).

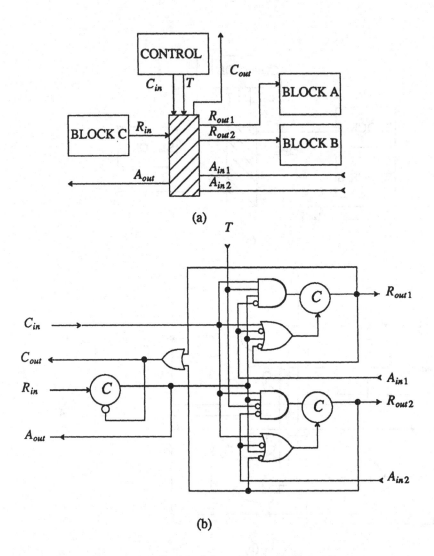

**Figure 3-22.** (a). A demultiplexer with two output destinations. (b). Logic implementation of the demultiplexer interconnection circuit specified by (3.9).

directly by a syntax-translation tool. We would like to have a convenient description for the complete system, rather than local specifications, as the front-end input to the synthesis program. Perhaps optimization can be included in this translation phase.

Our experience in synthesizing various interconnection circuits suggests that processor architecture plays an important role in specifying proper guarded commands. In the next chapter, we will describe the design of a self-timed programmable digital signal processor, and the related issues concerning the processor flow control, programmability, and processor architectures.

## REFERENCES

1.  E. W. Dijkstra, "Guarded Commands, Nondeterminacy and Formal Derivation of Programs," *Communications of the ACM* **18**(8) pp. 453-457 (Aug. 1975).

2.  T. A. Chu, "Synthesis of Self-Timed Control Circuits from Graphs: An Example," *Proc. IEEE 1986 ICCD*, pp. 565-571 (Oct. 1986).

3.  R. E. Miller, *Switching Theory*, John Wiley & Sons, Inc., New York (1965).

4.  F. Commoner and A. W. Holt, "Marked Directed Graphs," *Journal of Computer and System Sciences* **5** pp. 511-523 (1971).

5.  J. van de Snepscheut, "Trace Theory and VLSI Design," *Lecture Notes on Computer Science 200*, Springer-Verlag, (1985).

6.  J. C. Ebergen, "A formal Approach to Designing Delay-Insensitive Circuits," *Computer Science Notes*, Eindhoven University of Technology, (October 1988).

7.  A. J. Martin, "Compiling Communicating Processes into Delay-Insensitive VLSI Circuits," *Distributed Computing* **1** pp. 226-234 (1986).

8.  D. Misunas, "Petri Nets and Speed Independent Design," *Communications of ACM* **16**(8) pp. 474-481 (Aug. 1973).

9.  T. S. Balraj and M. J. Foster, "Miss Manners: A Specialized Silicon Compiler for Synchronizers," *Advanced Research in VLSI*, The MIT Press, (April 1986).

10. T. H.-Y. Meng, R. W. Brodersen, and D. G. Messerschmitt, "Automatic Synthesis of Asynchronous Circuits from High Level Specifications," *Submitted to IEEE Trans. on CAD*, (July 1987).

11. C. L. Seitz, "Self-Timed VLSI Systems," *Proc. of the Cal Tech Conference on VLSI*, (Jan. 1979).

12. D. E. Muller, "Infinite Sequences and Finite Machines," *Proc. 4th Annual IEEE Symposium on Switching Circuit Theory and Logical Design* **S-156** pp. 9-16 (Sept. 1963).

13. S. H. Unger, *Asynchronous Sequential Switching Circuits*, Wiley-Interscience, John Wiley & Sons, Inc., New York (1969).

14. A. J. Martin, "The Limitations to Delay-Insensitivity in Asynchronous Circuits," *Proc. of the Sixth MIT Conference in Advanced Research in VLSI*, pp. 263-278 (May 1990).

15. L. G. Heller and W. R. Griffin, "Cascode Voltage Switch Logic: A Differential CMOS Logic Family," *1984 IEEE ISSCC Digest of Technical Papers*, (Feb. 1984).

16. G. Jacobs and R. W. Brodersen, "Self-Timed Integrated Circuits for Digital Signal Processing Applications," *VLSI Signal Processing III*, IEEE PRESS, (November, 1988).

17. R. K. Brayton and C. McMullen, "Decomposition and Factorization of Boolean Expressions," *Proc. IEEE ICAS*, (May, 1982).

18. C. K. Erdelyi, W. R. Griffin, and R. D. Kilmoyer, "Cascode Voltage Switch Logic Design," *VLSI Design*, (October 1984).

19. T. H.-Y. Meng, R. W. Brodersen, and D. G. Messerschmitt, "Implementation of High Sampling Rate Adaptive Filters Using Asynchronous Design Techniquess," *VLSI Signal Processing III*, IEEE PRESS, (November, 1988).

20. D. E. Muller and W. S. Barsky, "A Theory of Asynchronous Circuits," *Proc. of International Symposium of the Theory of Switching*, pp. 204-243 Harvard University Press, (1959).

21. D. B. Gillies, "Flow Chart Notation for the Description of a Speed Independent Control," *Proc. 2nd Annual IEEE Symposium on Switching Circuit Theory and Logical Design* S-134(Oct. 1961).

22. R. E. Swartwout, "One Method for Designing Speed Independent Logic for a Control," *Proc. 2nd Annual IEEE Symposium on Switching Circuit Theory and Logical Design* S-134 (Oct. 1961).

23. R. E. Swartwout, "New Techniques for Designing Speed Independent Control Logic," *Proc. 5th Annual IEEE Symposium on Switching Circuit Theory and Logical Design* S-164(Nov. 1964).

24. T. Agerwala, "Putting Petri Nets to Work," *IEEE Computer*, pp. 85-94 (Dec. 1979).

25. J. B. Dennis, "Modular, Asynchronous Control Structure for a High Performance Processor," *Record of Project MAC Conf. Concurrent and Parallel Computation, ACM*, pp. 55-80 (1970).

26. S. S. Patil and J. B. Dennis, "The Description and Realization of Digital Systems," *IEEE COMPCON 72, Digest of Papers*, pp. 223-226 (1972).

27. J. L. Peterson, "Petri Nets," *Computing Surveys* 9(3) pp. 221-252 (Sept. 1977).

28. R. M. Keller, "Towards a Theory of Universal Speed-Independent Modules," *IEEE Trans. on Computers* C-23(1) pp. 21-33 (Jan. 1974).

29. A. S. Wojcik and K.-Y Fang, "On the Design of Three-Valued Asynchronous Modules," *IEEE Trans. on Computers* C-29(10)(Oct. 1980).

30. T. M. Carter, "ASSANSSIN: An Assembly, Specification and Analysis System for Speed Independent Control-Unit Design in Integrated Circuits Using PPL," *Master's Thesis*, Department of Computer Science, University of Utah, (June 1982).

31. D. L. Dill and E. M. Clarke, "Automatic Verification of Asynchronous Circuits Using Temporal Logic," *Proc. 1985 Chapel Hill Conference on Very Large Scale Integration*, pp. 127-143 Computer Science Press, (1985).

32. D. M. Chapiro, "Globally-Asynchronous Locally-Synchronous Systems," *Ph.D. Thesis*, (STAN-CS-1026)Stanford University, (Oct. 1986).

33. A. J. Martin, "The Design of a Self-Timed Circuit for Distributed Mutual Exclusion," *Proc. 1985 Chapel Hill Conference on Very Large Scale Integration*, pp. 245-283 Computer Science Press, (1985).

34. M. Hack, "Analysis of Production Schemata by Petri Nets," *MAC TR-94, Project MAC, MIT*, (Feb. 1972).

35. E. B. Eichelberger, "Hazard Detection in Combinational and Sequential Switching Circuits," *IBM Journal*, (March 1965).

36. C. Mead and L Conway, *Chap. 7, Introduction to VLSI Systems*, Addison-Wesley Publishing Company (1980).

37. S. Y. Kung and R. J. Gal-Ezer, "Synchronous versus Asynchronous Computation in Very Large Scale Integration Array Processors," *SPIE, Real Time Signal Processing V* 341(1982).

38. C. A. R. Hoare, "Communicating Sequential Processes," *Communications of the ACM* 21(8) pp. 666-677 (Aug. 1978).

39. B. C. Kuszmaul, "A Glitch in the Theory of Delay-Insensitive Circuits," *Proc. of ACM Workshop on Timing Issues in the Specification and Synthesis of Digital Systems (TAU '90)*, (August 1990).

40. R. L. Rudell and A. Sangiovanni-Vincentelli, "Multiple-Valued Minimization for PLA Optimization," *IEEE Trans. on CAD/ICAS*, (September 1987).

41. S. Burns, *Private Communications*, (May 1988).

42. A. J. Martin, *Private Communications*, (May 1988).

43. C. Berthet and E. Cerny, "An Algebraic Model for Asynchrnous Circuits Verifications," *Trans. on Computers* COM-37(7) pp. 835-847 (July 1988).

44. T. A. Chu, "A Method of Abstraction for Petri Nets," *International Workshop in Petri Nets and Performance Models*, (Aug. 1987).

# 4

---

# SELF-TIMED PROGRAMMABLE PROCESSORS

---

Designing a self-timed programmable processor is a challenging task because of its complexity. Despite the difficulties, self-timed programmable processors have recently drawn some attention from computer architecture designers [1,2,3]. The debate over which design discipline provides the most cost-effective system has not been resolved in four decades, and is technology-dependent. However, the interest in designing fully self-timed processors stems from a very different motivation. We like to experiment with the *average* speed exhibited by a self-timed programmable processor, since a moderate speed-up (two to five times faster) would have tremendous impact on how high performance digital systems will be implemented in the future.

This chapter describes the design considerations in implementing a self-timed programmable processor from an architectural description. To ensure a speed-independent design, processor architecture plays an important role in the design of interconnection circuits. We will discuss the issues relevant to the design of a programmable digital signal processor, such as pipelining, interconnection circuit specifications, data flow control, program flow control, feedback and initialization, I/O interface, and processor architectures. The system-level tradeoffs of using synchronous design vs. asynchronous design will be addressed and simulation results of an asynchronous version of a commercial digital signal processor will be given.

## 4.1. A PROGRAMMABLE SIGNAL PROCESSOR

In our asynchronous design model, we use a block diagram approach as opposed to the Petri-net discipline proposed by Dennis [4]. Block diagrams provide a simple picture of how a processor is to execute each instruction, and most designers are familiar with the notation. The modelling power of block diagrams, a subclass of Petri-nets, is sufficient for specifying the operations of programmable digital signal processors (PDSPs).

As outlined in Chapter 3, a self-timed processor is composed of two types of decoupled blocks: computation blocks and interconnection blocks. Shown in figure 4-1 is a self-timed programmable processor composed of a

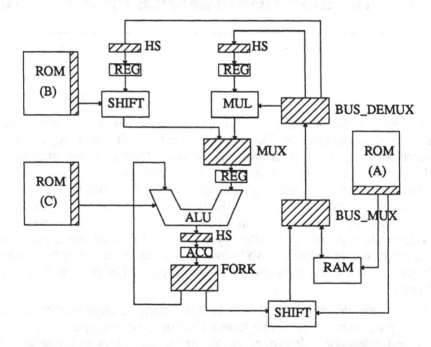

**Figure 4-1.** A self-timed programmable processor. Using the block diagram design approach, a processor architecture can be specified by decoupled computation blocks and interconnection blocks. The blocks labeled ALU, MUL, ROM, etc. are computation blocks and the blocks labeled MUX, FORK, etc. are interconnection blocks.

datapath, a data RAM, and three independent program ROM's. Aside from the three program ROM's, the architecture is representative of a common PDSP design.

The DCVSL computation blocks shown in figure 4-1, such as the arithmetic-logic-unit (ALU), the shifter (SHIFT), and the multiplier (MUL), have been designed and demonstrated a speed comparable to their synchronous counterparts [5,6]. The design of a 16×16 array multiplier (two output partial products) with completion signal generation will be described in Chapter 5 as an example of a DCVSL block. The circuitry for completion signal generation and handshake makes up only a small portion of the active area, in part because completion circuitry is only required at the last stage of the chain of DCVSL gates which make up the multiplier core.

An algorithm for synthesizing interconnection blocks with maximal possible concurrency under the hazard-free conditions was given in Chapter 3. All the interconnection circuits used in this chapter are synthesized by the mentioned automated tool.

## 4.2. DATA FLOW CONTROL

This section is mainly concerned with the design of self-timed data flow control given a processor architecture.

### 4.2.1. Data Latching Mechanism

The connection between computation blocks and interconnection blocks is governed by the request-complete signal pairs. A register is usually considered part of an interconnection block since it represent a pipeline delay in a data transfer. For example, in the full-handshake pipeline interconnection circuit shown in figure 3-15, the first C-element of the full-handshake controls the feedback acknowledge signal $A_{out}$, while the second C-element controls the request ($R_{out}$) to the next block. Intuitively, if a register is needed in a full-handshake circuit, it should be controlled by the acknowledge signal, as it indicates that the output data has been received by the succeeding block. In fact, for any pipeline connection using a full-handshake, $A_{out}$ is the only handshake signal that can be used for latching without corrupting either the input or the output data in the course of an unbounded-logic-delay type of operation.

In designing the data latching mechanism, we first decide whether a register is needed in an interconnection circuit. A register is necessary when there is a chance that the input to a DCVSL block changes while the DCVSL block is still evaluating the previous datum. A sufficient condition for this is when $R_{in}^{-} \cap (R_{out}^{+} \cap A_{in}^{-})$ is a possible condition in the signal transition graph,

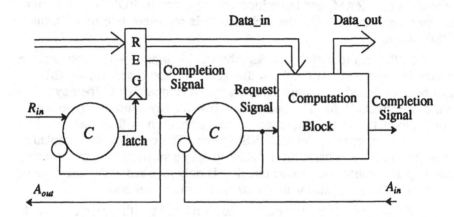

**Figure 4-2.** The connection between the pipeline register, the full-handshake circuit, and computation blocks.

which means that the output of the preceding DCVSL block has been precharged ($R_{in}^{-}$) but the succeeding DCVSL block has not finished evaluating ($R_{out}^{+} \cap A_{in}^{-}$). This condition is possible in both the full-handshake and the half-handshake, which is not surprising since there should be a pipeline register between any two pipeline stages. There are interconnection circuits in which such condition will never hold true, as in the non-pipeline interconnection circuit shown in Chapter 3.

The second task is to determine where to put the register, or which signal should be used to latch the data. From the behavioral description of the four handshake signals, the input data should be latched into the register after $R_{in}^{+} \cap A_{in}^{+}$ but before $A_{out}^{+} \cup R_{out}^{+}$: the first condition insures that valid data will be latched ($R_{in}^{+}$) only after the succeeding block has acknowledged the previous datum ($A_{in}^{+}$) and the second condition insures that valid data will be latched before acknowledged to the previous DCVSL block ($A_{out}^{+}$) and before fetched by the succeeding DCVSL block ($R_{out}^{+}$).

Any signal transition that happens within the above latching conditions can be used to control register latching. It follows that a full-handshake should use $A_{out}^{+}$ to latch data and a half-handshake should use $R_{out}^{+}$ to latch data. Deriving latching conditions is one of the most intriguing tasks in self-timed design. Different interconnection circuits may use different signals for latching, as depicted by the conditions described above. The above conditions were derived from the fact that DCVSL is used in computation blocks; other conditions may apply if other logic families were used. However, it is much easier to map the behavioral description to physical connection using signal transitions than circuit states. To illustrate how a behavioral description directs circuit synthesis, we will use figure 4-2 as an example.

In figure 4-2, a completion detection mechanism [7] is used to detect the completion of the register latching and to insure delay-insensitivity to the register latching operation. The completion signal is fed forward as an input to the second C-element and fed back as the acknowledge signal to the previous block. We use this completion signal instead of the output of the first C-element as the acknowledge signal $A_{out}$ to follow the behavioral description of $A_{out}$, which acknowledges the receipt of the current datum so that the preceding block can proceed with the next one. The output of the first C-element indicates only that the current datum has reached the handshake circuit, but the behavioral description of an *acknowledge* specifies that only after the datum has been *received* can $A_{out}$ go high. From previous discussion, there should be a register in a full-handshake and signal $A_{out}$ should be used for latching. The configuration in figure 4-2 is obtained by introducing a new signal, *latch*, into the signal transition graph. With $A_{out}^+$ replaced by $latch^+ \rightarrow A_{out}^+$ and $A_{out}^-$ replaced with $latch^- \rightarrow A_{out}^-$, the transition graph insures that only after the datum has been latched ($A_{out}^+$), will the next DCVSL block evaluate its input ($R_{out}^+$) and the preceding DCVSL block precharge ($R_{in}^-$). Synthesized directly from the modified signal transition graph with signal *latch*, the configuration shown in figure 4-2 is the latching mechanism used in each pipeline stage.

Since the DCVSL is a differential logic family, both true and complement data lines are needed at the input to DCVSL, but only true DCVSL outputs are transmitted to other DCVSL blocks or registers. Usually registers convert primary data bits to differential pairs. DCVSL generates a completion signal for each data bit; multiple completion signals are generated for multi-bit data. The interconnection circuit for multi-bit data can be treated as that required for multiple-input blocks, or the multiple completion signals can be passed through an AND gate.

## 4.2.2. Bus Control: Multiplexers and Demultiplexers

Chapter 3 illustrates the two most commonly used interconnection circuits for data flow control: the multiplexer and the demultiplexer. However, in self-timed design, since there is no global clock to regulate each operation, the completion of a data transfer must be properly acknowledged in order for the next instruction to be initiated. For example, the multiplexer as specified in Chapter 3 indicates that the unselected data, if there is one, should remain at the input of the multiplexer waiting to be fetched, as the corresponding acknowledge signal is not raised. This specification would do well for the multiplexer (MUX) connected to the multiplier (MUL) and shifter (SHIFT) blocks in figure 4-1.

On the other hand, the operation of the multiplexer BUS_MUX connected to RAM and SHIFT in figure 4-1 requires a different operation. The reason is that SHIFT delivers a new datum from FORK to BUS_MUX at each instruction cycle regardless of whether the datum will be selected or not; therefore datum needs to be flushed out in the next instruction if not selected. A simplified specification for BUS_MUX which transfers data from RAM and SHIFT to BUS_DEMUX can be expressed as

$$*[[R_{in1}^+ \cap C_{in}^+ \cap T \to R_{out}^+; A_{in}^+ \cap T \to A_{out1}^+, C_{out}^+] \parallel [R_{in2}^+ \cap T$$
$$\to A_{out2}^+] \mid R_{in2}^+ \cap C_{in}^+ \cap \bar{T} \to R_{out}^+; A_{in}^+ \cap \bar{T} \to A_{out1}^+, C_{out}^+], \qquad (4.1)$$

where a sequential construct is used in both alternative clauses to insure that there is no pipeline stage implied by the bus. In (4.1), $C_{in}$ is the request signal from the controller, $R_{in1}$ the request signal from RAM, and $R_{in2}$ the request signal from SHIFT. In the first alternative clause, the parallel construct $R_{in2}^+ \cap T \to A_{out2}^+$ is to indicate that if the datum from RAM is selected ($T$ high), then the bus must also issue an acknowledge signal fed back to SHIFT ($A_{out2}^+$) to prevent it from being deadlocked. The logic implementation of (4.1) is shown in figure 4-3. In general an interconnection circuit specification has to reflect the processor data flow strategy.

## 4.2.3. Memory Management

In our design model, memory is treated as an ordinary computation block, which in a datapath serves as either an input source (read operation) or an output destination (write operation). A computation block does not involve handshake operation other than the request and completion signals, but since a memory element (e.g. the RAM) performs two kinds of instructions (read and write), a read-write select needs to be controlled by a program ROM. The guarded command specifications for memory management depend on the type of memory elements (read or write exclusive, dual accesses, etc.) and on the architecture within which a memory element is embedded. As long as the read and write instruction sequence is determined by a controller, handshake operation for memory management can always be specified by guarded commands.

Consider a simple example of memory management in which the memory element RAM is read and write exclusive and controlled by a program ROM which issues a sequence of read or write instructions with direct addressing. Indirect addressing can be specified in a similar way. As shown in figure 4-4, the memory is simply a computation block with request signal $M_{req}$, read-write control signal $R/\bar{W}$, and completion signal $M_{comp}$. Whenever a read or write instruction is completed, $M_{comp}$ will go high to indicate the finish of an operation. Address and data lines are not shown in figure 4-4 for simplicity.

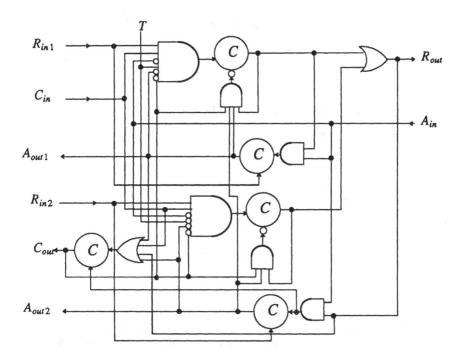

**Figure 4-3.** Logic implementation of BUS_MUX specified by the guard-ed command in (4.1).

The assumptions for a read instruction ($R/\overline{W}$ high) are that the address will be ready ($R_{in}^+$) at the input of the memory when $M_{req}$ is high and that the read data will be at the output of the memory when $M_{comp}$ is high. The assumptions for a write instruction ($R/\overline{W}$ low) are that both the address and the data will be ready ($R_{in}^+$ for both address and data) at the input of the memory when $M_{req}$ is high and that the instruction is completed when $M_{comp}$ is high. The assumption on $M_{req}$ is guaranteed by the interconnection block shown in figure 4-4, and the assumption on $M_{comp}$ relies on memory circuit design with completion signal generation. Completion signals of memory operation can be generated by accessing a dummy cell or by detecting an address change, as often used in high-performance static RAMs [8].

The interconnection circuit shown in figure 4-4 that uses a sequential operation for read instructions and a pipelined operation for write instructions can be specified as follows:

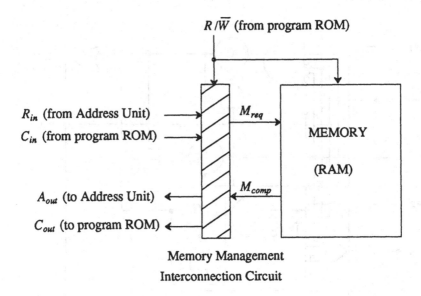

Figure 4-4. An example of memory management in which the memory element is read and write exclusive and controlled by a program ROM which issues a sequence of read/write instructions with direct addressing.

$$*[R_{in}^+ \cap C_{in}^+ \cap R/\overline{W} \rightarrow M_{req}^+; M_{comp}^+ \cap R/\overline{W} \rightarrow A_{out}^+; C_{in}^- \cap R_{in}^- \cap R/\overline{W}$$
$$\rightarrow M_{req}^- \mid R_{in}^+ \cap C_{in}^+ \cap R/\overline{\overline{W}} \rightarrow A_{out}^+; A_{out}^+ \cap R/\overline{\overline{W}} \rightarrow M_{req}^+], \quad (4.2)$$

where $C_{in}$ is the request signal from the program ROM to indicate when the read-write select $R/\overline{W}$ is valid, $C_{out}$ the corresponding acknowledge signal to the program ROM, $R_{in}$ the request signal for address and data, and $A_{out}$ the acknowledge signal of the completion of a memory operation. In (4.2), the write operation ($R/\overline{W}$ low) is straightforward but the read operation ($R/\overline{W}$ high) is a little bit tricky. Since we aim at sequential operation for read instruction, there is no pipeline register to latch the read data. When the read instruction is completed ($M_{comp}$ going high), it does not guarantee that the read data is fetched until $R_{in}$ goes low. Only after $R_{in}$ goes low can $M_{req}$ of the memory element be reset, as specified in (4.2). The circuit synthesized from (4.2) is shown in figure 4-5, which is the interconnection circuit used for the RAM in figure 4-1. To prevent the unfinished read-write instruction from being interrupted by the next instruction, the read-write select signal $R/\overline{W}$ must be stable between $C_{in}^+$ and $C_{out}^-$. If this environmental constraint is not satisfied, an internal read-write signal (a latched version of $R/\overline{W}$) will have to be devised. The circuit shown in figure 4-5 can be used as a simplified VME interface circuit upon which other operations in

VME can be added.

Multiple access contention can occur when more than one function unit request service of a memory element. In most PDSPs, memory is controlled statically, which means that multiple accesses are scheduled at compile time while application programs are being generated. In multiprocessor configurations, independent programs on different processors accessing a common global memory would require an arbiter. The design of correct arbiters can be found in Chapter 7 and other literature [9,10]. If an arbiter is necessary, the arbiter output can be connected to $R_{in}$ in the memory configuration of figure 4-4 to allow multiple accesses to the memory element.

### 4.2.4. Initialization and Feedback

A legal initial condition is a set of signal levels from which the circuit will not be driven into deadlock. A legal initial condition represents a steady state by holding the input signals at specific levels and the circuit can be correctly started by some invoking signal transitions. For example, the full-handshake circuit has eight legal initial conditions. Each initial condition represents a different state at which the circuit was presumably stopped and hence the computation must be restarted in a consistent way. With different initial conditions, the invoking signal transitions, which properly start up computation, are accordingly determined. Illustrated in table 4-1 are the

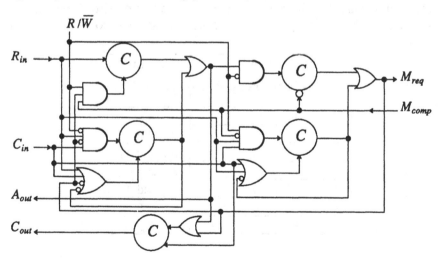

**Figure 4-5.** Logic implementation of the interconnection circuit specified in (4.2).

eight initial conditions and the corresponding invoking signal transitions for the full-handshake circuit. For simplicity we choose to use as the initial condition all the handshake signals being low, meaning that initially the circuit is waiting for new data to process.

Feedback such as the accumulator feedback around the ALU in figure 4-1 would result in deadlock if the initial condition is not proper. Since the accumulator is the only register in the loop, it will have to hold a valid output datum at the same time that it latches an input datum from the ALU. Therefore one more pipeline stage (a register and a pipeline interconnection circuit) must be added into the feedback loop to serve as a temporary buffer. As shown in figure 4-6(a), PI-A is an added pipeline stage of two input sources, one from the accumulator output and the other from an independent input block (represented by $R_{input}$). The added pipeline stage does not increase the number of pipeline delays in the loop, since with any set of legal initial conditions, only one sample delay can be specified in the loop. In general, to initiate $N$ pipeline delays in a feedback loop, at least $N+1$ pipeline states (registers and full-handshake circuits) are needed to realize a deadlock-free legal initial condition.

Legal initial conditions can be derived by inserting initial samples into the feedback loop. Initial samples can be realized by setting appropriate output request signals ($R_{out}$ in figure 4-2) high. In figure 4-6(a), since there are two pipeline stages in the loop and only one sample is to be initiated, the initial sample can be placed at the output of either of the two pipeline interconnection circuits. Shown in figure 4-6(b) is the initial condition if the initial

| Initial Conditions | | | | Invoking Transitions |
|:---:|:---:|:---:|:---:|:---:|
| $R_{in}$ | $A_{in}$ | $R_{out}$ | $A_{out}$ | |
| - | - | - | - | $R_{in}^+$ |
| - | + | - | + | $A_{in}^-$ |
| - | + | - | - | $R_{in}^+$ and $A_{in}^-$ |
| - | - | + | - | $R_{in}^+$ and $A_{in}^+$ |
| + | + | - | + | $R_{in}^-$ and $A_{in}^-$ |
| + | - | + | + | $R_{in}^-$ and $A_{in}^+$ |
| + | - | + | - | $A_{in}^+$ |
| + | + | + | + | $R_{in}^-$ |

**Table 4-1.** Legal initial states of a full-handshake circuit and the corresponding invoking signal transitions.

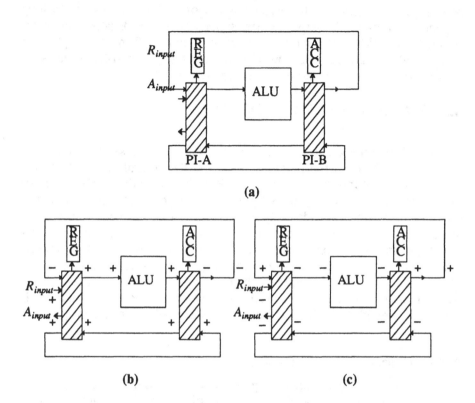

(a)

(b)                                    (c)

**Figure 4-6.** (a). A feedback interconnection configuration. Two registers (REG and ACC) are necessary to prevent the circuit from deadlock.
(b). The initial condition of the feedback loop if the sample delay is
placed at the output of PI-A. (c). The initial condition of the feedback
loop if the sample delay is placed at the output of PI-B.

sample is placed at the output of PI-A and in figure 4-6(c) the initial condition if the initial sample is placed at the output of PI-B. The circuit in figure
4-6(b) will be started by the falling edge of $R_{input}$ while the circuit in figure
4-6(c) by the rising edge of $R_{input}$. When a feedback module is connected to
the rest of the processor, consistent initial conditions must be maintained.
The fact that $R_{input}$ is initially high as in figure 4-6(b) will be propagated
through all the blocks preceding this feedback loop. We choose to use the
initial condition shown in figure 4-6(c) because it represents a simpler start-
up condition that is consistent with the earlier assertion that circuits should
be kept in the state of waiting for new data.

The number of initial samples that can be present within a feedback loop is
a function of the number of pipeline stages within the loop as well as of the
initial condition. At most $N$ samples can be inserted into a feedback loop of

$N+1$ pipeline stages, but the exact number of initial samples in the loop depends on how many output request signals of the handshake circuits are set high. If half-handshake circuits were used, only $N/2$ initial samples can be inserted into a feedback loop of $N$ registers and $N$ half-handshakes, as only alternate blocks can store valid samples.

## 4.3. PROGRAM FLOW CONTROL

Self-timed pipelining allows each pipeline stage to process data as soon as the data signals from the previous stage and the control signals from the corresponding controller (usually a ROM) are valid. The average throughput is dominated by the stage with the longest delay, but this longest delay is instruction dependent. Some instructions do not need to use every stage in a pipeline, for example the instruction of clearing the accumulator or the control instruction of setting a flag. Since processing is regulated by the availability of input data, a pipeline stage which is not used in a certain instruction can be asked to execute the next instruction, if of course its input data for the next instruction is available. The precedence of instructions is preserved through handshaking among the seemingly independent program ROMs.

### 4.3.1. Unconditional Branch

A branch instruction changes the program flow. Usually an unconditional branch is realized by connecting the output of the program ROM to its program counter (PC), as shown in figure 4-7. In figure 4-7, an unconditional branch is encoded by the control bit $L_{branch}$. The sequential interconnection circuit between the PC and the program ROM is for concurrent reset of both modules. The pipeline interconnection circuit at the output of the ROM is to implement the instruction pre-fetch mechanism usually engaged in a pipelined processor. If $L_{branch}$ is high, PC will load the next program address from the operand field of the ROM output and the program flow will be continued from the branched address. In this connection the PC functions as a register so that $L_{branch}$ and the branched address, latched by the register at the output of the ROM, will not affect the PC output until the request to the PC goes high again to calculate the next program address.

The feedback initialization scheme shown in figure 4-6(c) can be used to initialize the PC and ROM in figure 4-7. Since there is no second input source (except the $L_{branch}$ and the branched address) to the PC, $L_{branch}$ can be tested whenever the request signal to the PC is set high, as the request signal and $L_{branch}$ are in fact synchronized by the handshaking between the ROM and the PC.

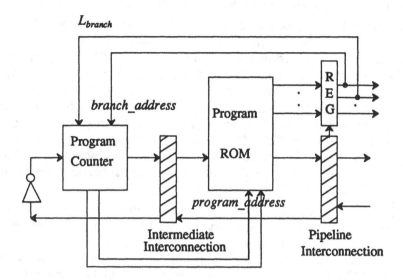

**Figure 4-7.** One possible realization of an unconditional branch instruction. If $L_{branch}$ is high, PC will load the next program address (the branched_address) from the operand field of the ROM output.

## 4.3.2. Conditional Branch

A conditional branch represents a special problem in the processor shown in figure 4-1, since each ROM operates independently except when there is an exchange of data. For example, if ROM (B) in figure 4-1 executes a conditional branch, the condition might not be available until ALU finishes its computation. Assume that the condition is tested at the output of the ALU. Then the ALU output must be transferred to ROM (B) in order to be tested for a possible branch. This data transfer requires handshaking between ROM (B) and ALU to ascertain instruction precedence, an operation to synchronize individual ROMs.

Shown in figure 4-8 is the schematic control flow among ALU, ROM (B) and ROM (C) (which controls the ALU). If the current instruction is a conditional branch, ROM (C) will issue a high on the control signal $C_{con\_branch}$, asking PI-3 to raise $R_{con\_branch}$ high after the ALU output is valid for testing. If the current instruction is a conditional branch, indicated by the output of the program ROM (B), PI-1 will latch the ALU output and test for the condition after $R_{con\_branch}$ goes high. Once a condition has been resolved, a feedback signal from PI-1 to the ALU will allow the ALU to proceed to the next instruction issued by ROM (C). A *con_flag* as a result will be latched and input to PC (B), similar to the $L_{branch}$ bit in an unconditional branch, to direct PC (B) to generate the appropriate next program address.

Handshake between pipeline stages in the datapath insures that the $C_{con\_branch}$ from ROM (C) will follow the operation in ROM (B) and that $R_{con\_branch}$ will be in synchrony with PI-1 to control PC (B). The purpose of the local feedback loop around ROM (B) and PC (B) is to ask PI-1 to check for $R_{con\_branch}$ if the current instruction is a conditional branch, so that $R_{con\_branch}$ would not be tested for other instructions. This feedback loop can be implemented in combination with the feedback loop from the ROM output to the PC for branched address loading, as mentioned in the unconditional branch. A similar control scheme is required between ALU, ROM (A), and ROM (C) if ROM (A) is to have a conditional branch based on the output of the ALU.

The conditional branching scheme shown in figure 4-8 provides a temporary dependency between ALU and ROM (B), which has no effect on other instructions not bearing such a precedent relation. A control bit (not shown in figure 4-8) and the branched address will be issued from ROM (B) to request PC (B) to test for the ALU output and to load the branched address if the condition holds. PI-3 could be either a pipeline interconnection or a sequential interconnection depending on whether the pre-fetch mechanism is preferred.

Asynchronous handshaking allows a form of reconfigurable datapaths for different instructions, at the expense of complicated interconnection circuitry. Hardware overhead in interconnection circuits and the delay penalty of driving many gates as a consequence of the many handshake signals involved in each instruction may render this approach impractical. For a simpler program control scheme, no-operations (NOPs) can be used to fill up those pipeline stages which do not have data to process while an "end of operation" is needed in order to continue the next instruction at every stage. With the facility of NOPs, the instruction sequence can be preserved with only pipeline handshakes. Of course we lose the flexibility of instruction-dependent datapaths but the hardware implementation is much simpler. For example, the three ROMs in figure 4-1 can be reduced to a single ROM with pipeline registers at its output to match corresponding pipeline stages. These considerations are discussed in the next section.

### 4.3.3. I/O Interface

The interface of a self-timed processor to the external synchronous world is usually through the use of synchronizers. Synchronizers, like arbiters, exhibit the metastable phenomenon if the sampling clock and the input signal make transitions at approximately the same time. As mentioned in Chapter 1, metastable circuits are not unique to self-timed systems; any system that facilitates non-deterministic operations such as fair mutually exclusive

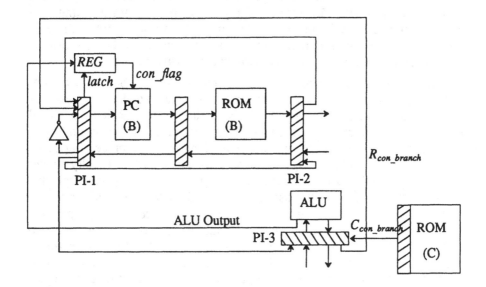

**Figure 4-8.** Schematic control flow of a conditional branch instruction. Handshaking between ALU and PC (B) is through $R_{con\_branch}$, which is to be raised high when the ALU output is valid and to be checked by PI-1 to insure instruction precedence. For simplicity, address and data lines are not shown in this figure. Except for $C_{con\_branch}$, $con\_flag$ and the ALU output, all other signals shown are handshake signals.

memory accesses would have to incorporate metastable circuits. For real-time applications, the effect of using non-deterministic operations is that the probability that the system would fail to respond to its input is nonzero, regardless of there being a global clock or not. On the other hand, if no metastable circuits are needed, a hard real-time constraint can be set up so that the worst-case data pattern be accommodated within the input sampling period (often many times of the instruction cycle time) and that synchronization failure will never occur.

Contrary to past expectations, combinational circuits built of differential logic for completion signal generation usually result in the worst-case delay for all input data patterns (if there is no carry-chain or the equivalent in the operation). The lack of data dependency simplifies the task of determining maximum sample rates.

In synchronous designs, the clock period is estimated by calculating the delay incurred in the global critical path, resulting in a constant worst-case cycle time for every instruction; in self-timed designs, processing cycle time can be estimated in the same way, but within a much smaller range of locality because only local interconnections need to be considered, and it is instruction-dependent. The approximately constant delay for all input data patterns incurred in self-timed computation blocks further simplies the estimation task, and the best performance of the worst-case data pattern, as necessitated by hard real-time constraint, can be easily obtained.

To calculate achievable input sampling rates for systems built of synchronous processors, one counts the number of instructions and multiplies it by the instruction cycle time, where the cycle time is regulated by a global clock. For systems built of self-timed processors, since the processing cycle time is instruction-dependent, we can achieve an *average* processing speed for a given *task*. This average processing time *per task* is usually higher than that of a globally clocked processor. The reason is evident from the fact that circuitry is by nature asynchronous and clocks are merely a means to attain synchronization points. For example, if the computation per task takes a constant computation time without using pipelines, in order to increase hardware efficiency, a synchronous processor would need to subdivide this computation time and finish the computation is a time interval which is a multiple integer of the clock cycle. Since the clock cycle represents the longest latency of all possible instructions, the overall computation time (the inverse of throughput) would be greater than the original constant time. If a self-timed programmable processor is used, the computation time per task is the original constant (possibly with some overhead introduced by handshaking), thus achieving a higher system throughput than that of the synchronous processor.

To use self-timed systems for real-time applications, we must insure that the worst-case computation time per sample be less than the sampling period. A simple model of a self-timed system interfacing to an A/D and a D/A converters is shown in figure 4-9. At the input end, an A/D converter is controlled by a sampling clock. The request signal $R_{in}$, indicating that a datum is valid at the output of the A/D converter, can be readily generated by the A/D converter. Since we assume that the input sampling period is longer than the total computation time for each sample, we do not need to connect $A_{out}$ back to the A/D converter. At the output end, a D/A converter is controlled by the same sampling clock. Registers at both the input and the output ends of the self-timed system are necessary to hold valid samples for one cycle time to prevent data from being corrupted due to an early precharge. Since the system throughput is regulated by the input sampling rate, sampling the register output at the same rate will provide correct output data. To complete the output handshake operation we simply connect $R_{out}$ to $A_{in}$.

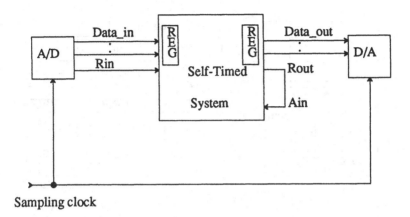

Sampling clock

**Figure 4-9.** A simple model of a self-timed system interfacing to the external synchronous world.

## 4.4. PROCESSOR ARCHITECTURE

Given that we can synthesize self-timed interconnection circuits from an architectural description, the next question is the best architecture for self-timed programmable processors. We are not at a point to answer this question yet, but some insights on the architectural issues were gained through our efforts of synthesizing self-timed processors based on established synchronous architectures [11]. Basically there are two types of design approaches, the data-stationary approach and the time-stationary approach. We will examine each in the following subsections.

### 4.4.1. Data-Stationary Architectures

In figure 4-10 a pipelined processor architecture programmed in a data-stationary way is illustrated. Each field of an instruction from the program ROM controls the operation of a single stage in the datapath, and is pipelined to match the corresponding pipeline stage in the datapath. Asynchrony allows each stage to start computation as soon as the data from the previous stage and the control signals from the ROM are valid, as indicated by the request signals from both blocks, and to issue a completion signal to the succeeding block and to the ROM as soon as it finishes its computation. The throughput is determined by the stage with the longest computation delay, but this longest delay and the stage it occurs in are instruction-dependent.

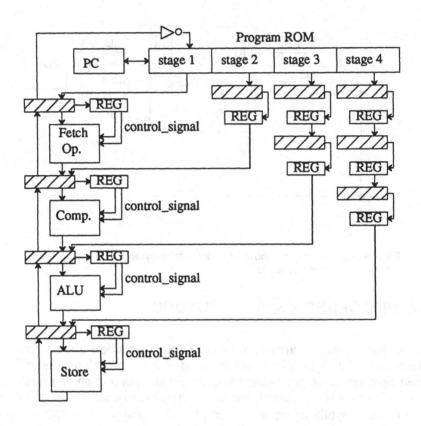

**Figure 4-10.** A pipelined processor architecture programmed in a data-stationary way.

Since operations are invoked by data flow, NOPs (no operations) must be inserted in those instruction fields which do not have any input data. A NOP can be realized by a simple pass transistor as shown in figure 4-11. Transistor M1 is controlled by the signal $C_{NOP}$ coming from the ROM. If the current instruction is a NOP, transistor M1 will be conducting, set by $C_{NOP}^{+}$, and $R_{in2}$ will be set high, independent of the state of the DCVSL computation block, as soon as the request signal $R_{out1}$ is high. Pass transistors are used so that a NOP is processed as fast as possible and will not constitute a limitation to processor throughput. As shown in figure 4-11, after $R_{in2}$ is high, $A_{out2}$ is fed back to PI-A to fetch the next instruction. However, the datum at the output of the DCVSL is corrupted since the computation is disrupted. Another pass transistor can be used to disable the latching of the corrupted datum into the input register at the next stage to avoid a loss of data.

NOPs can be seen as a sink operation to provide easy data flow control. With the facility of NOPs, any static instruction that involves a nonregular datapath can be executed efficiently. By a static instruction we mean that the program flow for the instruction is predetermined. A potential non-static, or dynamic, instruction is a conditional branch. As in the synchronous case, we can do two things about this: inserting NOPs before conditional branches to wait for the condition to be evaluated before starting the next instruction, or using some prediction policy and restoring processor states if the predicted path is not selected.

Conditional branches with inserted NOPs, which preserve the right sequence of instruction execution but introduce pipeline bubbles, can be readily realized by a branch bit and an address load operation into the program counter (PC) as depicted in figure 4-7, since both the condition and the branched address are available at the PC in the next instruction cycle.

Conditional branches with a prediction policy, on the other hand, need additional handshake operation between the condition flag and all the stages which will be affected by the decision made by the condition flag. For example, assume that the prediction policy is that the branching condition is not true. If the prediction is correct, operation continues without any change of contents in the control registers pre-fetched from the ROM. If prediction fails, this information should be fed back to load the PC with the branched address and to flush the contents of all the control registers shown in figure 4-10 with NOPs, in order to ensure that processor states will not be altered by incorrectly predicted instructions.

The fact that pass-transistor-NOPs should take no time to bypass computation blocks results in no specific advantages for conditional branches with a prediction policy, since self-timed pipeline bubbles, as implemented by NOP's, incur much shorter delays than clocked bubbles do in a synchronous design. In our design of a self-timed programmable processor no prediction policy is employed.

## 4.4.2. Time-Stationary Architectures

Figure 4-1 shows an example of a time-stationary architecture. ROM (A), (B), and (C) control the three pipeline stages and operate independently except when there is an exchange of data. The operation of each instruction fetched from each ROM is executed as soon as the input data are available.

Unlike data-stationary architectures, where the datapath is under sequential control, it is not practical for time-stationary architectures to employ a prediction policy in dealing with conditional branches. If the processor datapath has been changed according to an incorrect prediction, a complicated scheme of restoring processor states and a series of NOPs would have to be provided for such tasks.

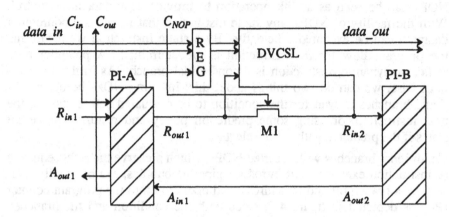

**Figure 4-11.** A NOP realized by a pass transistor.

As mentioned in the previous section, time-stationary processors allow a form of reconfigurable datapaths for different instructions. The question of the best architecture for a self-timed time-stationary processor is made more interesting by freedom to use instruction-dependent architectures. This approach displays the essence of self-timed programmable processor design, and we have shown that, through a synthesis procedure of self-timed interconnection circuits, the design of such processors is feasible and that it can be systematically synthesized from a description at the architectural level.

### 4.4.3. Multiple Function Units

In self-timed processor design, we can take advantage of the fact that request and acknowledge signals are available along the datapath. One possible architecture that cannot be easily implemented in synchronous design is the use of multiple function units in a pipelined datapath [12]. The reason is that the run-time scheduling for these multiple function units needs a much finer clock resolution if they are to be operated concurrently. For example, if a multiplier in a pipelined datapath incurs a much longer delay than other computation blocks and we do not want to further pipeline the multiplier to complicate programming, we can have two multipliers in parallel and leave to an interconnection circuit the task of routing and sequencing operands to the two multipliers. To the programmer, there is only one stage of a multipler with essentially twice the throughput (if two consecutive multiplication instructions are issued). The guarded command for the interconnection circuit at the input to multiple function units is simply a demultiplexer specified by (3.9) and at the output a multiplexer specified by (3.7). Both interconnection circuits are controlled by a select signal from the controller, and it is obvious that to obtain maximum throughput, alternating select signals should be used.

## 4.5. SIMULATION AND PERFORMANCE EVALUATION

In order to determine the performance which can be achieved in a self-timed design, the limitations on performance must be identified. There are considerable degrees of freedom to improve these limitations and some potential future directions will be discussed.

Although we base our logic synthesis procedure on a correct-by-construction principle, simulation is still required for a variety of purposes. Verification of proper operation of the complete processor is of primary importance. Timing analysis of delays due to actual circuit elements gives throughput information for different application programs and makes possible the evaluation of various processor architectures.

After the functionality of a system has been verified as correct, emulation is performed at the hardware level. Most conventional register-transfer level simulators cannot properly simulate self-timed computation since there is no global clock to determine when data is transferred. A time-queued event-driven [13,14] simulator capable of handling hazard detection and self-scheduling was therefore developed in Franz Lisp to facilitate accurate timing analysis and performance evaluations for systems built of asynchronous components.

### 4.5.1. Simulation of Processor Operations

Once a self-timed processor has been designed, simulation is required for performance evaluation, especially in programmable processors where the computation latency is instruction-dependent. We simulated the self-timed programmable processor in figure 4-1 with a simple FIR filter program. Each block in figure 4-1 is declared a node, which can be specified either by a behavioral description, such as an adder with a data-dependent delay, or by a logic network with more precise timing specifications. Our simulations used behavioral descriptions for computation blocks and logic descriptions for interconnection blocks, so that the processing cycle time for each instruction can be efficiently generated at the block level and that self-timed interconnection circuits can be correctly simulated at the gate level.

Information was gained on the speed-up over a synchronous version of this architecture, assuming that the computation blocks are of the same speed in both designs. A speed-up by a factor on the order of two or more was observed as a result of instruction dependencies. This is in contrast to the instruction rate of a synchronous processor, which is set by the slowest instruction. This instruction-dependent speed-up becomes more important as the variation of the computation times of the various instructions increases.

However, the most important advantage to self-timed circuits is the elimination of a global clock. The amount of improvement which will be achieved due to this simplification is dependent on processor architecture and actual circuit implementation.

## 4.5.2. Throughput of Pipelined Architecture

For a pipelined architecture, the system throughput is the reciprocal of the additive delay of the longest stage plus precharge plus one pipeline interconnection handshake (two set-reset cycles of a C-element if the full-handshake is used). The delays of interconnection handshake and the precharge can be considered as overhead as compared with a synchronous design, although in a synchronous design overhead such as non-overlap time will be needed to compensate for clock skew. This time-delay overhead ranges from 6 *nsec* to 15 *nsec* (in 1.6μ CMOS technology) depending on the data bit width and interconnection circuits used.

A nominal 10 *nsec* overhead for the computation of 16-bit data may seem too much to be practically acceptable. But this overhead will be reduced in direct proportion with the gate delay in more advanced technologies. We expect this overhead to be well below 2 *nsec* (for 16-bit data) in 0.8μ CMOS technology with better designs, which will allow a *system* throughput up to a few hundred Mhz to be possible without distributing a global clock [15].

Since only local communication sets the interconnection throughput limitation in a self-timed design, the system throughput of a pipelined architecture is independent of the system complexity if the delay of the longest stage remains unchanged. It is often argued that in synchronous design clock lines can be pipelined along with pipelined data, and hence the overall system throughput can be made independent of the length of the pipe. However, this pipelined clock scheme imposes severe limitations on the system configuration and thus is only useful in limited applications.

We have illustrated the feasibility and outlined the design procedure for a self-timed PDSP, through the design approach of separated computation and interconnection blocks and a procedure for automatic synthesis of self-timed interconnection circuits. Aside from fully custom designs, self-timed programmable processors offer intrinsic concurrency not available through a clocked design. The question of the best architecture for general-purpose self-timed computers can only be answered based on a complete understanding of the underlying interconnection design. Hopefully in the future, driven by performance requirements, the design of self-timed multiprocessor computers could be addressed which would allow all the benefits of technological advances to be exploited in the future supercomputer systems.

# REFERENCES

1.  A. J. Martin, *Private Communications*, (May 1988).

2.  M. T. Ilovich, "High Performance Programmable DSP Architectures," *Ph.D. Dissertation, UC Berkeley, ERL memo 88/31*, (June 1988).

3.  C. H. van Berkel and R. W. J. J. Saeijs, "Compilation of Communicating Processes into Delay-Insensitive Circuits," *Proceedings of ICCD 1988*, (October, 1988).

4.  J. B. Dennis, "Modular, Asynchronous Control Structure for a High Performance Processor," *Record of Project MAC Conf. Concurrent and Parallel Computation, ACM*, pp. 55-80 (1970).

5.  G. M. Jacobs and R. W. Brodersen, "Self-Timed Integrated Circuits for Digital Signal Processing Applications," *VLSI Signal Processing III*, IEEE PRESS, (November, 1988).

6.  T. H.-Y. Meng, R. W. Brodersen, and D. G. Messerschmitt, "A Clock-Free Chip Set for High Sampling Rate Adaptive Filters," *Journal of VLSI Signal Processing* 1 pp. 365-384 (April 1990).

7.  A. J. Martin, "Compiling Communicating Processes into Delay-Insensitive VLSI Circuits," *Distributed Computing* 1 pp. 226-234 (1986).

8.  S. E. Schuster, et al., "A 15ns CMOS 64K RAM," *1986 IEEE ISSCC Digest of Techinal Papers*, pp. 206-207 (February 1986).

9.  A. J. Martin, "The Design of a Self-Timed Circuit for Distributed Mutual Exclusion," *Proc. 1985 Chapel Hill Conference on VLSI*, pp. 245-283 Computer Science Press, (1985).

10.  D. L. Dill and E. M. Clarke, "Automatic Verification of Asynchronous Circuits Using Temporal Logic," *Proc. 1985 Chapel Hill Conference on VLSI*, pp. 127-143 Computer Science Press, (1985).

11.  E. A. Lee, "Programmable DSP Architecture," *ASSP Magazines*, (Dec. 1988 and Jan. 1989).

12.  W.-M. Hwu and Y. N. Patt, "Checkpoint Repair for High-Performance Out-of-Order Execution Machines," *Trans. on Computers* C-36(12)(Dec. 1987).

13.  A. R. Newton, "Timing, Logic, and Mixed-Mode Simulation for Large MOS Integrated Circuits," *Computer Design Aids for VLSI Circuit*, pp. 175-240 Matinus Nijhoff Publishers, (1984).

14.  S. A. Szygenda and E. W. Thompson, "Modeling and Digital Simulation for Design Verification and Diagnosis," *IEEE Trans. on Computers* C-

25(12) pp. 1242-1253 (Dec 1976).

15. T. E. Williams, "An Overhead-free Self-timed Division Chip," *Stanford Technical Report*, (August 1990).

# 5

---

# A CHIP SET FOR ADAPTIVE FILTERS

---

The application of self-timed design to high performance digital signal processing systems is one promising approach to alleviating the difficulty of global clock synchronization. To demonstrate this approach for a typical signal processing task, the system architecture and circuit design of a chip set for implementing high rate adaptive lattice filters is described. The first part of this chapter discusses the performance considerations involved in the design of self-timed circuits through an actual IC implementation of such a chip set [1]. The second part of this chapter demonstrates the ease with which self-timed systems can be designed with a minimum amount of design efforts. The experimental implementation showed that self-timed design simplifies the design process since no effort was devoted to clock distribution, nor were any timing simulations to verify clock skew necessary.

In conventional synchronous design, the throughput of a parallel processing system can be limited by the global constraint of clock distribution, which in turn limits the physical size of the system, the clock rate, or both. While in self-timed design, localized forward-only connection allows computation to be extended and sped up using pipelining without any global constraint on the overall system throughput. A pipelined architecture can thus achieve an *arbitrary* high rate consistent with input/output bandwidth limitations (such as the speed of A/D converters), as will be demonstrated in the filter application in this chapter.

## 5.1. PROPERTIES OF SELF-TIMED DESIGNS

With the ability to automatically synthesize self-timed interconnection circuits [2], we believe that self-timed processing chips and their assembly onto boards is a considerably simpler task than designing processors which require global synchronization. We will now discuss the design of self-timed systems with emphasis on the performance issue.

### 5.1.1. Properties of Computation Blocks

DCVSL computation blocks use the so-called *dual–rail* [3] coded signals inside the logic, which compute both the data signal and its complement. Since both rising and falling transitions have to be completed before the completion signal can go high, if there is no carry-chain in the design, DCVSL computation blocks usually give the same worst-case computation delay independent of input data patterns. As mentioned in Chapter 4, this lack of data dependency to the first order is actually an advantage for real-time signal processing, since the worst case computation delay can be easily determined.

Once a self-timed processor has been designed, there is no way to slow down the internal computation. The throughput can be controlled by issuing the request signal at a certain rate, but the computation within the processor is performed at the full speed. This property has an impact on testing: unlike synchronous circuits, we cannot slow down the circuitry to make it work with a slower *clock*.

In self-timed design it is tempting to introduce shortcuts that improve performance at the expense of speed-independence. However, these shortcuts necessitate extensive timing simulations to verify correctness under varying conditions. To minimize design complexity we therefore generally stayed true to the speed independence design, with the result that timing simulations were required only to predict performance and not to verify correctness.

### 5.1.2. Properties of Interconnection Blocks

When computation blocks communicate between one another, an interconnection block is designed to control the data transfer mechanism. We require that an interconnection circuit be speed-independent; i.e., the circuit's behavior is independent of the relative gate delays among all the physical elements in the circuit. The automated synthesis procedure in

Chapter 3 has been used to generate all the interconnection circuits required in the adaptive filter chip set. Since the synthesis procedure uses a deterministic algorithm to maximize circuit concurrency and minimize Boolean expressions, the interconnection circuits synthesized usually require much less hardware complexity than those synthesized from a collection of compiled modules [4,5]. However, the time-delay overhead incurred by the circuit delays cannot be ignored, and will be discussed in Section 5.4 as part of the chip performance evaluation.

### 5.1.3. Speed-Independence

We have discussed the connection of registers to interconnection circuits in Chapter 4. For the convenience of our discussion here, the connection of a full-handshake circuit to the corresponding register is replicated in figure 5-1. The handshake circuit uses the rising edge of output acknowledge signal $A_{out}$ to control register latching. A detailed description on how to derive latching signals for different interconnection circuits was given in Chapter 4. Now we take a closer look at the latch. The register shown in figure 5-1 is an edge-triggered latch. The reason that an edge-triggered latch is needed instead of a level-triggered one is for fast acknowledge signal feedback. Notice that in figure 5-1, $R_{out}$ (request signal) is fedback to the first C-element without waiting for the completion of the computation block connected to it. If a level-triggered latch were used, the feedback signal to the first C-element will have to wait for the completion of the succeeding computation block to prevent an early precharge of its input data [6]. It can be

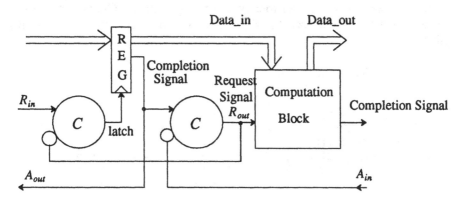

**Figure 5-1.** The connection between the pipeline register, the pipeline handshake circuit, and computation blocks.

easily verified that if a feedback signal is generated after the completion of the succeeding block, the handshake circuit would enforce that only alternate blocks can compute at the same time and that the hardware utilization is reduced to at most 50%. An edge-triggered latch does not constitute a practical problem in our design since it is strictly a local operation. The latch can be designed such that the operation is not sensitive to the slope of the triggering signal.

In order to insure the speed-independence of the latching operation, a completion detection for register latching is shown in figure 5-2. The logic compares the input and output data bits. If the latching signal is high and the input and the output data bits are identical, the completion signal goes high. During the reset cycle, a low on the latching signal will immediately reset the completion signal and therefore incurs only a short delay for resetting. After the data bits have been latched and stable, the completion signal from the register is fed forward as an input to the second C-element. We could match the delays of register latching and the second C-element to gain some performance improvement. However, simulation showed that any attempts at gate delay matching proved to be unreliable, and thus we chose to stay true with a speed-independent design.

The data latching mechanism shown in figure 5-1 is gate-delay-insensitive (speed-independent) but not wire-delay-insensitive, as the wire delays of data lines may be different from that of the request signal $R_{in}$. If the request signal and data lines do not match in wire delays, differential pairs of data lines, or coded data lines, will have to be transmitted and the completion signal $R_{in}$ will be generated locally at the input to the C-element. In our design, we also assume that the wire delays *within* each logic block is negligible.

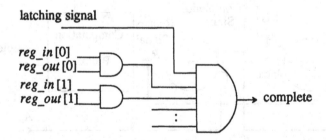

**Figure 5-2.** The completion detection mechanism for register latching. The logic compares the input and output data bits. If the latching signal is high and the input and the output data bits are identical, the completion signal goes high.

## 5.1.4. System Initialization

A self-timed system can be initialized in several states, depending on whether initial samples are necessary or not (see Chapter 4). For a feed-forward configuration, the initial samples do not have any impact on output behavior, so we might as well set all the initial conditions (the request and acknowledge signals controlled by interconnection blocks) to zero. For a feedback configuration, a nonzero number of samples within the feedback loop are necessary to avoid deadlock, and a sample can be represented by setting the corresponding request signal high in a pipeline interconnection circuit. The output of a memory element, for example a C-element, can be easily set high with some initialization circuitry. But when combined with DCVSL computation blocks, the initialization process is slightly more complicated than just setting C-elements.

When the request signal of a DCVSL block is set high while its input register is being cleared at initialization, the output data of the DCVSL will be undefined, which may result in a garbage sample that will propagate through the feedback loop. This is no problem for data samples in our particular adaptive filtering application, since the effect of an arbitrary initial sample will fade away because of the adaptation procedure. On the other hand, an undefined logic state may deadlock the system right at the start. A two-phase initialization is adopted in our design to cope with this problem.

In the first initialization phase the precharge signal to each DCVSL block is reset to zero to force a clean datum for the succeeding block, and then in the second phase those request signals corresponding to a sample delay in a feedback loop are set high so that a clean completion signal can be subsequently generated. In our design, a *set* memory element can be set high by

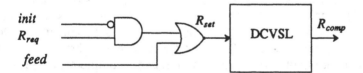

**Figure 5-3.** The logic used to allow a two-phase initialization. During the first phase, *init* is high and all the reset memory elements and register contents are reset to low. During the second phase, *feed* is set high and the request signal ($R_{set}$) corresponding to a sample delay in a feedback loop is set high through the OR-gate. A completion signal is subsequently generated by the DCVSL ($R_{comp}$), indicating that there is a sample in the DCVSL block. Both *init* and *feed* can then be reset and the system can be started without deadlock.

an *init* signal and a *reset* memory element can be reset low by the same *init* signal, but set memory elements and reset memory elements are disjoint. A simple logic network as shown in figure 5-3 is used to realize the two-phase initialization. During the first phase, when *init* is high and *feed* is low, the logic output $R_{set}$ is low (precharge), and all the output data lines are set to a specific logic level. During the second phase, when *feed* is set high, the logic output $R_{set}$ is high and a completion signal ($R_{comp}$) can be generated by the DCVSL. indicating that there is a sample in the DCVSL block. Both *init* and *feed* can then be reset and the system can be started without deadlock.

## 5.2. A SELF-TIMED ARRAY MULTIPLIER

In this section we will describe the design of a self-timed array multiplier to illustrate the circuit design considerations often encountered in designing DCVSL computation blocks and the solutions to these problems.

### 5.2.1. Architecture of the Multiplier

A multiplication in adaptive filters is usually followed by an accumulation operation, and the final addition in the multiplication can be combined with the accumulation to save the delay time of one carry-propagate add. As shown in figure 5-4, the multiplier core outputs two partial products to be added in the succeeding accumulate-adder, which consists of one carry-save add and one final carry-propagate add. The multiplier core is composed of three basic cells: registers, Booth encoders, and carry-save adders. The 16 multiplier digits are grouped into eight groups and each group is Booth encoded to either shift or negate the multiplicand [7]. Eight partial products are computed and added using a modified Wallace tree structure [8]. The architecture of the multiplier core shown in figure 5-4 consists of eight Booth encoders operating concurrently and six carry-save adders with four of them operating sequentially. The *inv* signals from encoders to carry-save adders implement the plus-one operation in a two's complement representation of a negative number. The maximum quantization error is designed to be less than one least-significant bit by adding a 0.5 to one of the carry-save adders for rounding. The multiplier core occupies a silicon area of 2.7mm×3mm in a 1.6μm CMOS design, and a computational latency of 40 *ns* was estimated from simulation using *irsim* [9] with the 1.6μm parameters. The test results of the design will be given in the next section.

To increase the system throughput, a pipeline stage was added between the multiplier core and the accumulate-adder. We could have deep-pipelined the multiplier core, because pipelined computation within a feedback loop can be compensated by look-ahead computation without changing the

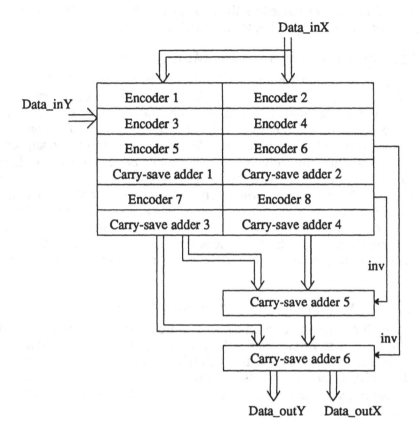

**Figure 5-4.** The architecture of the multiplier core, which consists of eight Booth encoders operating concurrently and six carry-save adders with four of them operating sequentially.

system's input-output behavior [10,11]. However, since the computational delay of the multiplier core has been reduced to only one Booth encoding plus four carry-save adds for 16-bit data, the hardware and delay-time overhead incurred in adding more pipeline stages, such as the circuitry for pipeline registers and the delay time for register latching, completion detection, and handshake operation, rules out the practicability of having one more stage of pipelining within the multiplier core.

The accumulate-adder consists of a carry-save adder at the front and a carry-propagate adder at the end to add three input operands per cycle; two of the three operands come from the output partial products of the multiplier core, with the third from a second source. The accumulate-adder occupies a silicon area of 1.1mm×1mm in a 1.6$\mu m$ CMOS design. A propagation delay of 30 *ns* was estimated by simulation using the 1.6μm parameters and the test results will be given in the next section.

## 5.2.2. Circuit Design for the Multiplier Core

The goal in designing the multiplier core was to reduce the DCVSL delay by allowing maximal concurrent operation in both precharge and evaluation phases. Layout is another important design factor since a Wallace tree structure consumes more routing area than sequential adds. Data bits have to be shifted two places between each Booth encoder array to be aligned at the proper position for addition. Since each data bit is coded in a differential form, four metal lines (two for the sum and two for the carry) have to be shifted two places per array, which constitute on the order of a 40% routing area overhead as compared to similar structures in a synchronous design. In this subsection, the basic cells used in the multiplier core, modified from a design provided by Jacobs & Brodersen [12], will be given and the strategies of obtaining minimum precharge and completion generation overhead will be discussed.

### The Booth Encoder and the Carry-Save Adder

For simplicity the logic schematic of a Booth encoder, rather than its DCVSL gate, is shown in figure 5-5. The Booth encoder encodes three consecutive bits ($Y0, Y1, Y2$) of the multiplier into three different operations on the multiplicand: $1X$ stands for shift 0, $2X$ stands for shift 1, and *inv* stands for negate. The DCVSL gates for a Booth encoder require more than

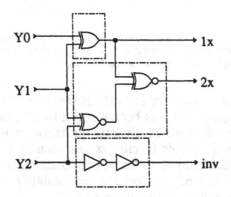

**Figure 5-5.** The logic schematic of a Booth encoder, which encodes three consecutive bits ($Y0, Y1, Y2$) of the multiplier into three different operations on the multiplicand: $1x$ stands for shift 0, $2x$ stands for shift 1, and *inv* stands for negate.

20 transistors in the NMOS trees because of the complex selection combinations. However, DCVSL shows performance advantages as compared to static logic gates when logic becomes complex, since essentially only one gate delay is incurred for each logic function.

The output of a Booth encoder is a partial product (a shifted and/or negated copy of the multiplicand) to be added with other partial products. The DCVSL gates for implementing a carry-save adder are shown in figure 5-6. Modifications on the transistor sizes for better driving capability and on the PMOS transistors to precharge every node high to prevent charge-sharing were manually tailored for the specific needs of this multiplier design. As shown in figure 5-6, a carry-save adder takes three input bits $A$, $B$, and $C$ and adds them together to produce two output bits $CARRY$ and $SUM$. Since there is no carry propagate between digits, a carry-save adder performs an addition in one gate delay.

Eight rows of Booth encoders and four rows of carry-save adders are divided into two planes and connected by abutment, and the last two rows of carry-save adders are connected by channel route. The die photo of the

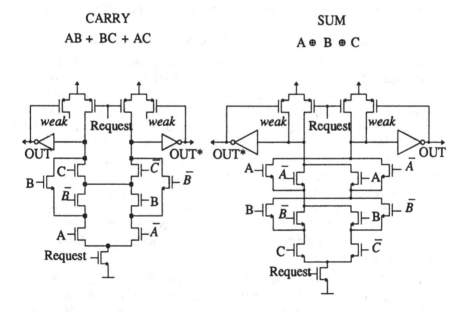

Figure 5-6. The circuit schematic of a DCVSL 6arry-save adder. Every internal node is precharged high during the precharge phase to avoid the charge sharing problem. A carry-save add consists of two separate NMOS trees, a carry tree and a sum tree. The differential output lines are fed into the next row of carry-save adds.

multiplier core is shown in figure 5-14, where eight rows of Booth encoders and four rows of carry-save adders are divided into two planes and connected by abutment, and the last two rows of carry-save adders are connected by channel routing. Routing for two data shifts as mentioned earlier was done by two layers of metal lines and one layer of polysilicon, occupying approximately 50% of the layout. Since there is no carry-chain in the multiplier design, the computational delay of the multiplier is independent of input data patterns.

## Concurrent Precharge

Even though the five rows of sequential DCVSL gates (one for the Booth encoding and the other four for carry-save adds) evaluate data sequentially, they can be precharged concurrently. The precharge signal ($R_{out}^-$) is driven through three levels of distributed buffers to drive more than 320 PMOS precharge transistors at the same time. When the outputs of the last row of the DCVSL gates are reset to low (the inverse of a precharged high), the multiplier completion signal will go low, indicating the precharge has been completed. By using the concurrent precharge scheme, a 16-bit multiplier precharge operation can be finished within 6 $ns$ in a 1.6$\mu m$ CMOS design.

## Concurrent Request In Sequentially Connected DCVSL

We had assumed that when the request is high, the input data to the corresponding DCVSL gate must be valid and stable. This assumption is violated by a concurrent request since the DCVSL gates are connected sequentially. For example, when the request is set high, the input data to the second row of the DCVSL gates, which are the outputs of the first row of the DCVSL gates, are definitely not valid yet. But since every output bit of DCVSL is reset to low in precharge, it is guaranteed that when the request is high at a DCVSL gate, the input bits to that gate can only go from low to high, but never from high to low. Since DCVSL consists of only NMOS logic trees, input gate voltages going from low to high will only increase the conductivity of the NMOS tree but never the reverse; therefore the one directional transition property is preserved, similar to the operation of Domino dynamic circuits.

If all the transistors along an NMOS tree are driven high, the output line will eventually go high, through an inverter at the top of the NMOS tree, indicating the completion of the evaluation for that particular tree. Valid data travel through multiple DCVSL gates sequentially, but the request signal for all the DCVSL gates are set high at the same time. After every differential pair of the last row reaches a high at one of the differential output data lines, a completion signal for the whole DCVSL block can be generated. The concurrent request scheme takes advantage of the asynchronous nature of circuit delays in order to obtain the shortest overall computational delay during the evaluation phase. Since completion signal generation can be viewed as a synchronization point, the number of unnecessary completion detections should be minimized. A completion signal is only generated at the output of the last row of a DCVSL block.

## Completion Signal Generation

The completion signal for every output data pair of a DCVSL gate can be generated through an OR gate with inputs connected to the differential output data pair. As shown in figure 5-7, the completion signal for multi-bit data can be generated by a C-element tree which detects the completion of every data pair. In our design, two-level AND gates were used to detect the 16 completion signals at the output of a DCVSL gate, based on the assumption that precharge delay is approximately a constant for all data bits.

To reduce the overhead incurred by completion signal generation, we can parallel computation and completion signal generation, using the *pipelined completion detection* scheme [6]. In this scheme, level-triggered latches would have been necessary to latch both the primary and complementary data pairs. Level-triggered latches will require the acknowledge signal to be fedback only after the completion of the succeeding computation block. The overall performance gain of using pipelined completion compared to the edge-triggered latching mechanism shown in figure 5-1 will depend on the actual circuit functionality. In general, simple iterative computation will benefit from using the pipelined completion detection scheme, as the

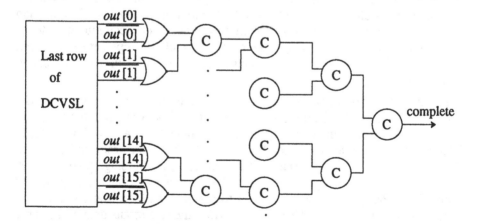

**Figure 5-7.** The completion detection for a 16-bit DCVSL block. C-elements can be replaced by simple AND gates if precharge delay is approximately constant for all data bits.

overhead introduced by low hardware efficiency could be more than compensated by avoiding the delay for completion signal generation.

Another obvious alternative to the completion signal generation would be to look at the NMOS tree which is expected to display the longest delay. This scheme usually works fine if there is no carry chain in the computation, as in this case each NMOS tree would exhibit the same amount of computational delay independent of input data patterns. The tradeoffs between adopting a speed-independent design and using a compromised scheme for better performance are among the most interesting issues raised for future study.

## 5.2.3. Circuit Design for the Accumulate-adder

The accumulate-adder is a much simpler design than the multiplier since only one carry-save adder and one carry-propagate adder are needed. Subtraction is realized by an *inv* bit to control a simple DCVSL tree which selects either the primary data or their complements at the output of input registers and feeds the selected data to a row of carry-save adders. The outputs of the carry-save adders are fed directly into a row of carry-propagate adders for computation of one single sum out of three input operands.

The DCVSL block of a carry-save adder is given in figure 5-6. A carry-propagate adder is implemented by feeding the carry-bit of a carry-save adder to one of the three inputs of the next carry-save adder. The concurrent precharge and request plays an important role here; if a sequential precharge and request scheme had been used, one 16-bit carry-propagate adder would have incurred 16 sequential precharge delays and 16 sequential requests and completional detections, which would constitute a total delay of more than $200\,ns$ in $1.6\mu m$ CMOS! The die photo of the 16-bit accumulate-adder is shown in figure 5-14; the first three rows of the circuit represent three input registers, the fourth row is a carry-save adder, and the fifth row is a carry-propagate adder.

The relative computational delays as exhibited by this carry-propagate adder are rather interesting. The DCVSL carry gate is designed such that the carry will go high immediately after two of the input bits ($A$ and $B$) are high, regardless of the level of the third ($C$). If the carry from the previous DCVSL stage is taken as input $C$ to the present DCVSL stage, two 1's on $A$ and $B$ will issue a carry to the next DCVSL stage without waiting for the carry from the previous stage, similar to the design of the Manchester carry chain. Consequently two 1's or two 0's from the two inputs data at the same bit position will cut the carry-chain into two independent chains, which means that the two carry chains can then compute concurrently. From the viewpoint of data-dependent computational delays, the more the concurrent 1's and 0's the faster the carry-propagate will be.

## 5.3.  A VECTORIZED ADAPTIVE LATTICE FILTER

We chose to demonstrate the design of a self-timed DSP system using the vectorized adaptive filter scheme introduced in [13]. The vectorized adaptive filter allows arbitrarily high sampling rates for a given speed of hardware, at the expense of parallel hardware. This allow us to test the highest throughput achievable through a self-timed design. We chose to use a lattice filter [14] in our design because the adaptation feedback in a lattice filter can be limited to a single stage. This results in much less computation within the feedback path and higher inherent speed in a given technology. The successive stages of a lattice filter are independent and hence can be pipelined.

### 5.3.1.  The Filter Architecture

Linear adaptive filtering is a special case of a first-order linear dynamical system. The adaptation of a single stage of a lattice filter structure can be described by two equations, the state-update equation

$$k(T) = a(T)k(T-1) + b(T) , \qquad (5.1)$$

and the output computation

$$y(T) = f(k(T), T) , \qquad (5.2)$$

where $a(T)$ and $b(T)$ are read-out (memoryless) functions of the input signals at time $T$, and $y(T)$ is a read-out function of both the state $k(T)$ and the input signals at time $T$. Since the computation of $y(T)$ is memoryless, the calculation can be pipeline-interleaved [15] and the computation throughput can be increased without a theoretical limit, if input signals and state information can be provided at a sufficiently high speed. However, the state-update represents a recursive calculation which results in an iteration bound on the system throughput [16]. In order to relax the iteration bound without changing the system's dynamical properties, (5.1) can be written as

$$
\begin{aligned}
k(T+L-1) &= a(T+L-1)a(T+L-2)\cdots a(T+1)a(T)k(T-1) \\
&+ a(T+L-1)a(T+L-2)\cdots a(T+1)b(T) + \cdots \\
&+ a(T+L-1)b(T+L-2) + b(T+L-1) \\
&= c(T,L)k(T-1) + d(T,L)
\end{aligned} \qquad (5.3)
$$

where $c(T,L)$ and $d(T,L)$ are memoryless functions of past input signals, but independent of the state $k(T)$. $c(T,L)$ and $d(T,L)$ can be calculated with high throughput using pipelining and parallelism, and the recursion $k(T+L-1) = c(T,L)k(T-1) + d(T,L)$ needs only be computed every $L$ samples. Therefore the iteration bound is increased by a factor of $L$. $L$ is

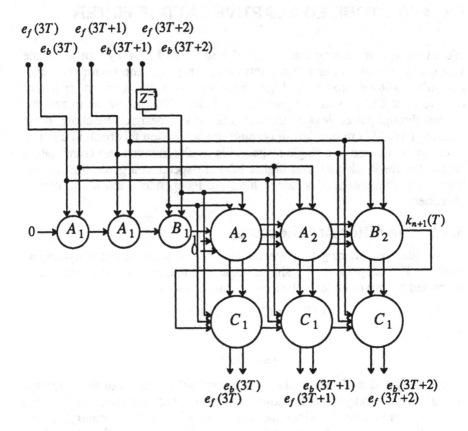

**Figure 5-8.** An LMS adaptive lattice filter stage with a vector size of three.

called the *vector size*, since a vector of $L$ samples will be processed concurrently.

A normalized least-mean-squares (LMS) or stochastic gradient adaptation algorithm is chosen for our example because of its simplicity and the absence of numerical overflow compared to transversal recursive least-squares algorithms. A vectorized LMS lattice filter stage with $L = 3$ is shown in figure 5-8 and the operations that each processor needs to perform are shown in figure 5-9. The derivation of the algorithm for normalized LMS lattice filters can be found in [5]. Processors $A_1$ and $A_2$ jointly calculate the $c(T,L)$ and $d(T,L)$ in (5.3), processor $B_1$ and $B_2$ calculate the state-update $k(T)$, and processor $C_1$ performs the output computation $y(T)$.

For every processing cycle, a vector of $L$ input samples are processed and a vector of $L$ output samples are calculated. The processing speed is $L$ times

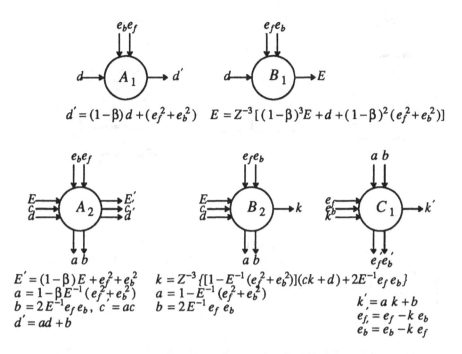

**Figure 5-9.** Operations required of each processor shown in figure 5-8.

slower than the filter sampling rate. Since there is no theoretical limit to the number of samples allowed in a vector, the filter sampling rate is not constrained by the processing speed, but rather it is limited only by the I/O bandwidth (such as the speed of A/D converters). However, the higher sampling rate is achieved at the expenses of additional hardware complexity and system latency, and very high sampling rates will require multi-chip realizations. For example, at the sampling rate of 100MHz the computational requirement is approximately 3 billion operations per second per lattice stage. Since there is no global clock routing at the board level in a self-timed design, pipelined computation hardware can be easily extended without any degradation in overall throughput as both the vector size and the number of lattice filter stages are increased.

Our goal is to design a set of chips that can be connected at the board level to achieve any sampling rate consistent with I/O limitations. This requires the partitioning of a single lattice filter stage into a set of chips which can be replicated to achieve any vector size. In this partitioning we attempted to minimize the number of different chips that had to be designed, allow flexibility in the vector size (ie. amount of parallelism), and minimize the data bandwidth (number of pins) on each chip.

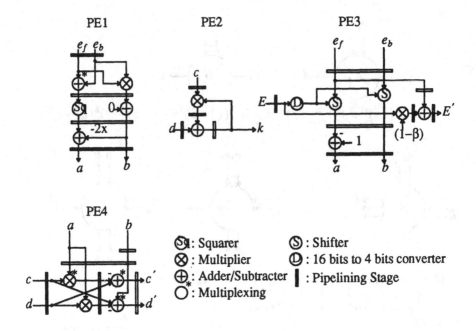

**Figure 5-10.** A chip set designed to implement the LMS adaptive lattice filter shown in figure 5-8. The partition was done in such a way that any required filter sampling rate can be achieved by replicating these chips at the board level without redesign. PE1 and PE4 have built-in hardware multiplexers so that the chip function can be controlled by external signals to form a reconfigurable datapath.

We found a partitioning that uses five different chips and meets these requirements. Block diagrams of four of them are shown in figure 5-10, and a fifth chip simply implements variable-length pipeline stages. Two of the four chips (PE1 and PE4) have built-in hardware multiplexers so that the chip function can be controlled by external signals to form a reconfigurable datapath; otherwise eight different chip designs would have been required. The same chip set can also be used to construct a lattice LMS joint process estimator. These chips can be replicated as required and interconnected at the board level to achieve any desired system sampling rate.

The architecture of computation chips and pipeline stages for a vectorized LMS lattice filter with vector size $L = 3$ is shown in figure 5-11. Pipeline registers, indicated by rectangles in figure 5-11, are the most commonly required circuit elements. Since the number of pipeline stages preceding different computation chips ranges from 1 to $3L + 7$, one pipeline chip was designed which has a variable length and can be cascaded.

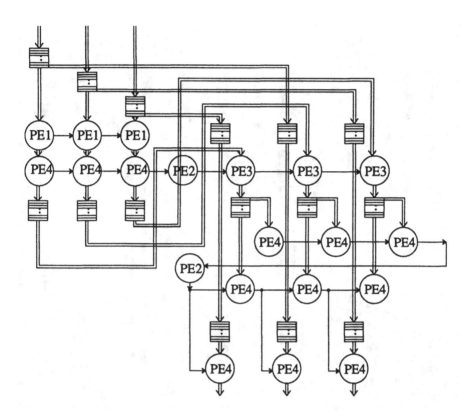

**Figure 5-11.** The connections among computation chips and pipeline registers for a vectorized LMS lattice filter stage with the vector size equal to three.

## 5.3.2. The Pipeline Chip

In the pipeline chip, each pipeline register is controlled by a pipeline handshake circuit and the number of pipeline stages can be specified by external signals. Because the signals in a normalized lattice filter always come in pairs, two variable-length pipeline stages are included in one pipeline chip. The die photo of a pipeline chip of 108 pins with 4 I/O paths (two for input and two for output) is shown in figure 5-12, where each I/O path consists of 18 bits This chip contains two chains of pipeline stages where each chain can be programmed by external signals to act as a variable-length pipeline for up to 16 stages. The chip size is 6.4mm×7.2mm (including pads) in a 1.6 $\mu m$ CMOS design. The estimated delay from pin-in to pin-out is 30 $ns$ from simulation using the 1.6 $\mu m$ parameters. This delay is considerable and due to the large capacitance on the common bus lines, each

**Figure 5-12.** The die photo of a pipeline chip with two variable-length pipeline stages, and with four I/O paths where each I/O path consists of 18 bits (108 pins total).

of which is connected to seven tri-state buffer diffusions. Fortunately this delay has no impact on the system throughput, as in the pipeline architecture the throughput is dominated by the pipeline stage with the longest delay, which in our case is the multiplication stage.

### 5.3.3. The Computation Chips

The four computation chips shown in figure 5-10 were designed by interconnecting various computation (multipliers, shifters, etc.) and interconnection (full-handshake, multiplexing, etc.) macrocells using an automatic place and route tool.

#### PE1: Input Signal Energy Estimates

PE1 performs the computation of estimating the input signal energies, and computes two quantities:

$$a = e_f^2 + e_b^2$$
$$b = e_f\, e_b\,.$$

(5.4)

To reduce the numbers of multipliers in the calculation of (5.4), the actual operations are carried out by

$$b = e_f\, e_b$$
$$a = (e_f + e_b)^2 - 2b,$$

(5.5)

where one extra addition is introduced to save one multiplication. The * sign at the adder in this chip (shown in figure 5-10) represents a multiplexing point, where the other possible configuration of this chip is simply multiplexing $e_b$ with zero to implement the computation of $a = e_c^2$ necessary in an LMS joint process estimator (in this case, $e_f$ is fed with $e_c$, the residual error of the joint process estimator). The die photo of PE1 is shown in figure 5-13, which occupies a silicon area of $5.8mm \times 7.2mm$ (including pads) with 28.7K transistors using $1.6\mu m$ CMOS technology.

#### PE2: Feedback Computation

PE2 computes the state-update recursion:

$$k(T) = c(T)k(T-L) + d(T).$$

(5.6)

This feedback computation chip is the simplest in the sense that only one multiplier and one accumulator are implemented on a single chip; however, the operation is the most complicated in initialization. Since there are two pipeline stages within the feedback loop, one between the multiplier and the accumulator and the other at the output of the accumulator, two initial samples must be introduced into the feedback loop. As described in the previous section, the set logic shown in figure 5-3 is used to artificially initialize the two samples into the two pipeline registers. The longest pipeline delay along this feedback loop determines the system throughput, as the handshake stage with the two-phase initialization logic incurs more run-time overhead than usual handshake stages. PE2 was implemented with 13.5K transistors. The die photo of PE2 is shown in figure 5-14, which occupies a

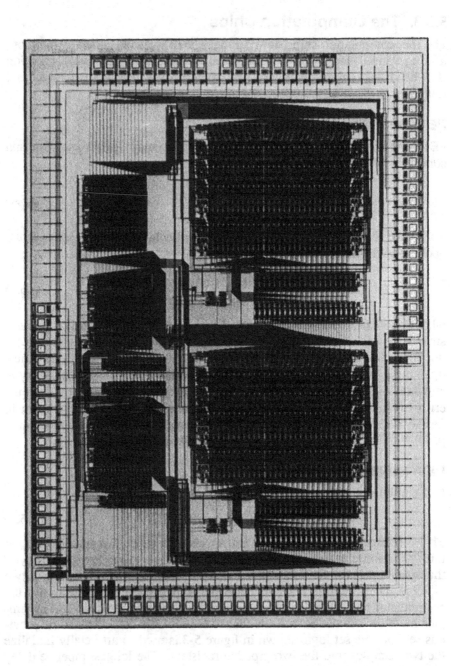

**Figure 5-13.** The die photo of chip PE1.

silicon area of approximately $4mm \times 3mm$ using $1.6\mu m$ CMOS technology

## PE3: Energy Normalization Computation

PE3 performs the normalization computation in a normalized LMS algorithm. The computation can be described by the following equations:

$$a' = 1 - \frac{a}{E}$$

$$b' = \frac{b}{E} \qquad (5.7)$$

$$E' = (1-\beta)E + a,$$

where $a$ and $b$ are input signal energy estimates, $a'$ and $b'$ are normalized input signal energy estimates, and $E$ and $E'$ are the present and the next energy norms. $\beta$ controls the adaptation speed of the norm, and therefore the convergence characteristics of the algorithm; a detailed discussion of how to choose the value of $\beta$ can be found in [5]. The division necessary in energy normalization computation is approximated by a shift operator. The 16-bit energy norm $E$ is first rounded to its nearest neighbor in a 4-bit representation, and these 4 bits are then used to control a shifter to perform the divide. Simulations showed that this approximation of using shifters for dividers will not cause much degradation in convergence properties if an LMS algorithm is used [17]; the reason is that the energy norm term will be exactly canceled out in the division if the input signal is stationary. The operation of PE3 is implemented by a multiply and accumulate chip (PE4) in combination with a shifter chip provided by Jacobs and Brodersen [12].

## PE4: Output Computation

PE4 performs three different calculations depending on how the multiplexers are configured, as listed below:

1. Filter residual error computation:

$$c' = d - a\,c$$

$$d' = c - d\,a. \qquad (5.8)$$

2. State look-ahead computation:

$$c' = a\,c$$

$$d' = a\,d + b. \qquad (5.9)$$

3. State-update computation:

$$c' = a\,c + b. \qquad (5.10)$$

The multiplexers were designed such that both the feed-forward request

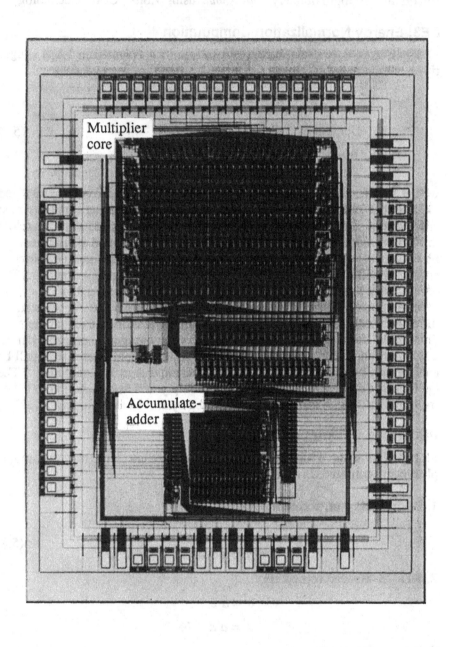

**Figure 5-14.** The die photo of PE2.

signals and the feedback acknowledge signals are properly selected. Run-time penalty is paid for having these multiplexing operations, but three different chips would have been necessary if multiplexers were not used. Since the processing cycle is determined by the feedback loop in PE2, the multiplexing overhead as introduced here will not degrade the overall system performance. PE4 occupies a silicon area of $7.6mm \times 7.6mm$ (including pads) with 33.2K transistors using a $1.6\mu m$ CMOS design. The die photo of this chip is shown in figure 5-15.

**Figure 5-15.** The die photo of PE4.

## 5.4. PERFORMANCE EVALUATION

### 5.4.1. Testing Circuitry

Scan-path registers can replace pipeline registers for testing purposes. By holding interconnection signals in a steady state, test vectors can be shifted in and out of scan-path registers within self-timed processors as is typically done in synchronous testing, and the internal states of a self-timed processor can be both observable and controllable.

In the pipeline chip, the output data bits at every pipeline stage is observable through external control; therefore no testing circuitry is necessary. In the computation chips, since we designed DCVSL computation blocks with registers combined in the layout to save chip area, adding extra scan-path registers into the datapath would complicate the floor plan and increase the run-time delay. Since completion signals are derived directly from DCVSL output data lines, internal completion signals are connected to pinouts instead for testing purposes. If a fabrication error happened along the datapath, the completion signal would stay low because DCVSL outputs are dual-rail coded. The DCVSL macrocell in which the error resides can thus be easily identified. Functional errors, if any, can be caught through input test vectors.

### 5.4.2. Chip Performance Evaluations

The chip set was fabricated through MOSIS in 1.6μm CMOS technology. The chips were tested using the Tektronic DAS9100 testing system with run-time pattern generation and data acquisition. The test results are summerized in table 5-1. The computational delay for the 16×16-bit multiplier (two output partial products) and the 16-bit accumulator (three input operands) are comparable to their synchronous counterparts. However, substantial overhead, compared to *perfect* synchronous design, is paid for the self-timed design in the handshake delay, precharge, and completion signal generation.

Handshake delays are on the order of a couple gate delays. Delays for precharge and completion detection are usually a function of the data bit width. In our design, since we follow the strictest speed-independent rule, the operations of precharge, handshake, computation, and completion detection are processed sequentially. Even under this extreme conservative condition the processing rate can be as high as 16Mhz for one 16×16 multiplication.

The feasibility of designing self-timed circuits with higher performance than their synchronous counterparts at the *chip level* is an issue of current debate. To answer this question, more investigation has to be done in exploiting all

| 16-bit Macro | Multiplier | Accum. | Register | C-element |
|---|---|---|---|---|
| Computation/Set | 30 ns | 20 ns | 4 ns | 4 ns |
| Precharge/Reset | 6 ns | 6 ns | - | 1 ns |
| Completion Sig. | 8 ns | 8 ns | - | - |

**Table 5-1.** The test results of the self-timed computation chips using DCVSL and pipeline handshake circuits. The computation latencies are comparable to their synchronous counterparts, but the overhead paid for precharge and completion signal generation is substantial.

possible parallelism in the self-timed design regime. For example, the delay incurred by completion signal generation can be reduced [18], as discussed in the previous section. Synchronous design has been studied for more than three decades, and is now a mature technology. The work presented here is the first attempt to try to assess the advantages and disadvantages of using self-timed circuits for DSP applications by actually implementing these chips. As will be discussed in the next subsection, the modular design approach as promised by using self-timed components is expected to give performance advantages at the *board level*, even though an individual chip may not function at the same speed as a perfectly matched synchronous chip.

## 5.4.3. Board Level Interconnection

Systems built using a clock-free approach can be easily extended without problems in global synchronization. This is particularly important for our vectorized lattice filter application, where we want to use the same chip set to achieve a variety of sampling rates (depending on application) at the board level.

In our vectorized lattice filter application, a lower throughput, say as a result of self-timed design or slower technology, can be compensated by a larger vector size for a constant filter sampling rate. The total chip count for the computation chips increases linearly with the vector size, while the chip count for the pipeline chips increases quadratically. The number of chips of each type required for a 100MHz LMS lattice filter stage is given in table 5-2. As one benchmark, for a vector size of 10, 60% of the chip count is for pipelining. This overhead is due to the fact that on the order of $L^2$ data samples are stored within the system, which is a consequence of trading latency for throughput.

The large number of chips needed for a vectorized lattice filter was caused in part by the fact that the chip set was designed for programmable rate

| Cycle Time | Vector Size | Pipe. Chip | PE1 | PE2 | PE3 | PE4 | Total Counts |
|---|---|---|---|---|---|---|---|
| 20ns | 2 | 8 | 2 | 2 | 2 | 7 | 21 |
| 50ns | 5 | 30 | 5 | 2 | 5 | 19 | 61 |
| 80ns | 8 | 55 | 8 | 2 | 8 | 31 | 104 |
| 100ns | 10 | 89 | 10 | 2 | 10 | 39 | 150 |
| 200ns | 20 | 221 | 20 | 2 | 20 | 79 | 342 |

**Table 5-2.** A summary of the various numbers of chips required for a 100MHz LMS lattice filter stage using the chip set in figure 5-10. Over 50% of the chip count is dedicated to pipeline registers.

adaptive filters. In order to make the filter sampling rate parameterizable at the board level, the computation partition was constrained by the three criteria as mentioned in Section 5.3.1, which leads to the chip set and chip counts shown in table 5-2. If the design were targeted on a fixed rate, all the pipeline stages can be implemented on the computation chips, which will reduce the chip count by at least 50%, but the filter sampling rate would then be dependent on the fabrication process. To accommodate process variations, design that exceeds specs is generally adopted, which would increase the design effort, and sometimes even the design cost.

The board level design in which the same chip set is used to achieve different filter sampling rates can be accomplished by an automated netlist generation, without worrying about clock distribution, and the desired sampling rate can always be achieved with a sufficiently large vector size. Communication latencies at the board and backplane level are not of concern because they are included within pipeline stages, and are automatically compensated by the handshaking. Building different systems with varying sampling rates using the chip set and this design approach should therefore require little design effort. On the other hand, a lower hardware cost for a given sampling rate at the expense of a considerably higher design cost could be obtained in current technologies using a synchronous design methodology, or even with a self-timed design dedicated to a given rate.

In this chapter, the design modularity which can be achieved with an self-timed design approach was demonstrated by the presentation of an architecture and chip partitioning for an adaptive lattice filter stage at an arbitrarily high sampling rate. The conservative speed-independent design strategy that we have adopted simplifies the design process at the expense of additional overhead in the interconnection circuits.

The relative performance of synchronous and self-timed design is highly technology dependent. At the feature sizes used in our design, a synchronous design would undoubtably have higher throughput at the chip level (there will be a crossover at some smaller feature size). However, from the perspective of ease of system design and performance at the system level, a self-timed design may be a winner. What is needed in the future is the aggressive pursuit and improvement of both synchronous and asynchronous design methodologies, and a comparison of their performance and design difficulty when applied to common applications using common technologies.

## REFERENCES

1.  T. H.-Y. Meng, R. W. Brodersen, and D. G. Messerschmitt, "Implementation of High Sampling Rate Adaptive Filters Using Asynchronous Design Techniquess," *VLSI Signal Processing III*, IEEE PRESS, (November, 1988).

2.  T. H.-Y. Meng, R. W. Brodersen, and D. G. Messerschmitt, "Asynchronous Logic Synthesis for Signal Processing from High Level Specifications," *IEEE ICCAD 87 Digest of Technical Papers*, (Nov. 1987).

3.  D. B. Armstrong, A. D. Friedman, and P. R. Manon, "Design of Asynchronous Circuits Assuming Unbounded Gate Delays," *IEEE Trans. on Computers* C-18(12)(Dec. 1969).

4.  D. Misunas, "Petri Nets and Speed Independent Design," *Communications of ACM* 16(8) pp. 474-481 (Aug. 1973).

5.  C.H. van Berkel and R. W.J.J. Saeijs, M. J. Honig, and D. G. Messerschmitt, "Adaptive Filters: Structures, Algorithms, and Applications," *Proceedings of ICCD 1988*, Kluwer Academic Publishers, (1984).

6.  T. E. Williams, M. Horowitz, R. L. Alverson, and T. S. Yang, "A Self-Timed Chip for Division," *Advanced Research in VLSI, Proc of 1987 Stanford Conference*, pp. 75-96 (March 1987).

7.  A. D. Booth, "A Signed Binary Multiplication Technique," *Q. J. Mech. Appl. Math.* 4(2) pp. 236-240 (1951).

8.  M. Santoro and M. Horowitz, "A Pipelined Iterative Array Multiplier," *IEEE ISSCC 88 Digest of Technical Papers*, (February, 1988).

9.  M. Horowitz, *IRSIM User's Manual*, Stanford University, (1988).

10. K. K. Parhi, "Pipelined VLSI Recursive Filter Architectures Using Scattered Look-Ahead and Decomposition," *Proceedings of ICASSP 1988*, (April, 1988).

11. K. K. Parhi and M. Hatamian, "A High Sampling Rate Recursive Digital Filter Chip," *VLSI Signal Processing III*, IEEE PRESS, (November, 1988).

12.   G. M. Jacobs and R. W. Brodersen, "Self-Timed Integrated Circuits for Digital Signal Processing Applications," *VLSI Signal Processing III*, IEEE PRESS, (November, 1988).

13.   T. H.-Y. Meng and D. G. Messerschmitt, "Arbitrarily High Sampling Rate Adaptive Filters," *IEEE Trans. on ASSP* **ASSP-35**(4)(April 1987).

14.   M. J. Shensa, "Recursive Least-Squares Lattice Algorithms: A Geometrical Approach," *IEEE Trans. Automatic Control* **AC-26** pp. 695-702 (June 1981).

15.   H.-H. Lu, E. A. Lee, and D. G. Messerschmitt, "Fast Recursive Filtering with Multiple Slow Processing Elements," *IEEE Trans. on CAS* **CAS-32**(11)(November, 1985).

16.   M. Renfors and Y. Neuvo, "The Maximum Sampling Rate of Digital Filters Under Hardware Speed Constraints," *IEEE Trans. on CAS* **CAS-28**(3)(March, 1981).

17.   W.-L. Chen, *Private Communications*, (1987).

18.   T. E. Williams, "An Overhead-free Self-timed Division Chip," *Stanford Technical Report*, (August 1990).

# 6

# ISOCHRONOUS INTERCONNECT

David G. Messerschmitt
Department of Electrical Engineering and Computer Sciences
University of California at Berkeley

Thus far in this book, self-timed circuits for synchronization of a digital system have been emphasized. This is an anisochronous technique, in which the operations in the digital system are ordered according to the completion of previous operations rather than being slaved to a clock. This is a promising approach for future scaled integrated circuit technologies, where interconnect delays tend to become large relative to circuit speeds, because the correct operation can be made delay independent. However, self-timed circuits are not the only way to achieve delay independence. In this chapter we consider a class of isochronous interconnect approaches, which include the synchronous design commonly used in the past as a special case [1,2]. These include, in addition to synchronous interconnect, mesochronous, pleisochronous, and heterochronous approaches. All these techniques are suggested by similar approaches that have long found application in digital communication, a domain where interconnect delays are routinely equal to many clock cycles [3].

In addition to describing these techniques, we determine some simple bounds on the throughput of both isochronous and anisochronous interconnect, and compare them. These bounds show a greater potential throughput for isochronous interconnect in the presence of large interconnect delays, basicly because of the elimination of the round-trip delay for the handshaking signals. Thus, isochronous techniques show promise of achieving high throughputs in practice as well as in theory, and this is borne out in the field of digital communication where multi-gigabit rates are commonly achieved using these isochronous approaches. In the course of considering these bounds, we uncover an approach to the unified treatment of pipelining in digital systems, one which treats computational and interconnect delays in a unified fashion.

## 6.1. SYNCHRONOUS INTERCONNECT

The most common design approach in the past has been synchronous interconnection, so we consider this first and then extend to other related approaches. As shown in figure 6-1, each element (or perhaps module) is provided a clock, as well as one or more signals that were generated with transitions slaved to the clock. The common clock controls the order of operations, insuring correct and reliable operation of the system. Since all operations are slaved to the clock, all signals in the system are isochronous.

We will first examine some fundamental limitations on the operation of synchronous interconnection, making idealistic assumptions about the ability to

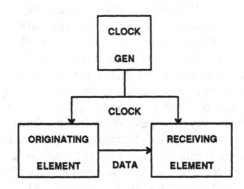

**Figure 6-1.** Synchronous interconnection, in which a common clock is used to synchronize computational elements.

control the clock phases in the system and neglecting interconnect delays. We will then show how pipeline registers can be used to extend the performance of synchronous interconnect, and make more realistic estimates of performance considering the effects of the inevitable variations in clock phase and interconnect delays. This will uncover the phenomenon of *clock skew*, which is a significant practical problem in synchronous interconnect in the presence of large interconnect delays.

## 6.1.1. Principle of Synchronous Interconnect

The fundamental principle of synchronous interconnection is illustrated in figure 6-2. In figure 6-2(a) a *computational block* C1 is connected to a *synchronizing register* R1 at its input. This register is *clocked* using the positive transitions of a periodic clock signal, where the assumption is that the output signal of the register changes synchronously with the positive transition of the clock. The computational block performs the same computation repeatedly on new input signals applied at each clock transition. The

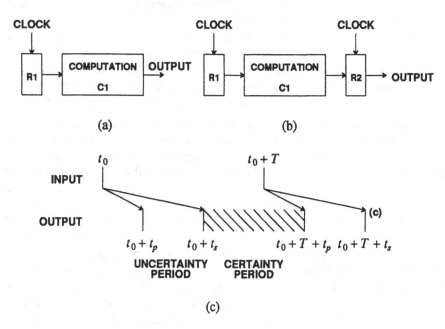

(a)                            (b)

(c)

**Figure 6-2.** Synchronizing a computation C1 by latching its input. (a) A register R1 at the input controls the starting time for the computation. (b) A second register R2 at the output samples the output signal during the certainty period. (c) Timing diagram showing the uncertainty and certainty period (crosshatched) as a function of the setting time $t_s$, the propagation time $t_p$, and the clock period $T$.

purpose of R1 is to control the time at which the computational block starts to perform its work, in order to synchronize it to other computational blocks in the system. This is an *edge-triggered* logic model, which we employ for its relative simplicity. There is also a more complicated *level-sensitive* model that leads to virtually identical conclusions.

The performance measures of interest in figure 6-2 are the *throughput* (rate at which the computation is repeated) and *computational latency* (delay from the time a new input is applied until the result is available). Focusing on the latter, inevitably the computational latency is not entirely predictable. It is likely that the output signal will change more than once before it finally settles to its final correct value. For example, if the output signal actually consists of $M > 1$ Boolean signals in parallel, as is often the case, some of those Boolean signals may transition before others, or some may transition more than once before reaching a steady-state value. This behavior is an inevitable consequence of real circuit implementation of the computational block, and presents a considerable problem in the synchronization to other computational blocks. Assume the computational block has a minimum time before *any* outputs transition, called the *propagation time* $t_p$, and a maximum time before all the outputs reach their final and correct values, called the *settling time* $t_s$. Since settling time is the minimum time before the result is guaranteed to be available, it is also the computational latency. It is assumed that the propagation and settling times can be characterized and insured over expected processing variations in the circuit fabrication.

The synchronous interconnect isolates the system behavior from these realities of circuit implementation by setting the clock period $T$ so that there is a *certainty period* during which the output signal is guaranteed to be correct, and then samples the output signal again (using another register R2) during this certainty period as shown in figure 6-2b.. Of course the phase of the clock for R2 must be adjusted to fall in this certainty period. This interconnect is called synchronous because the proper phase of clock to use at R2 must be known in advance — there cannot be any substantial uncertainty or time variation in this phase. With synchronous interconnect, the undesired behavior (multiple signal transitions and uncertain completion time) is hidden at the output of R2. We can then abstract the operation of the computational block, as viewed from the output of R2, as an element that completes its computation precisely at the positive transitions of the second clock, and this makes it easy to synchronize this block with others since the uncertainty period has been eliminated.

Given the propagation and settling times, the certainty period is shown as the crosshatched period in figure 6-2(c). The region of time not crosshatched, during which the signal may possibly be changing, is known as the *uncertainty period*. If two successive clock transitions come at times

$t_0$ and $(t_0 + T)$, the certainty period starts at time $(t_0 + t_s)$, the time when the steady-state output due to the computation starting at time $t_0$ is guaranteed, and ends at time $(t_0 + T + t_p)$, which is the earliest that the output can transition due to the new computation starting at time $(t_0 + T)$. The length of the certainty period is $(T + t_p - t_s)$. An acceptable register clocking phase for R2 requires that this period must be positive in length, or

$$T > t_s - t_p . \tag{6.1}$$

Alternatively, we can define the *throughput* of computational block C1 as the reciprocal of the clock period, and note that this throughput is upper-bounded by

$$\frac{1}{T} < \frac{1}{t_s - t_p} . \tag{6.2}$$

The length of the uncertainty period $(t_s - t_p)$ is a fundamental property of the computational block, and the maximum throughput is the reciprocal of this length. In contrast, the length of the certainty period depends on the clock period $T$, and the goal is generally to make this length as small as practical by choosing $T$ small.

The maximum throughput is dependent only on the length of the uncertainty period, $(t_s - t_p)$, and not directly on the settling time $t_s$. In figure 6-3 an example is given for throughput much higher than the reciprocal of the settling time (because the uncertainty time is a small fraction of the settling time). In figure 6-3, before each computation is completed, two more computations are initiated. At any point in time, there are three concurrent computations in progress.

The number of concurrent computations is limited only by the inevitable uncertainty in the computational latency. We can give three examples that illustrate a range of possibilities.

**Example 6-1.**
  Consider a fiber optic digital communication system, where the "computational block" is not a computation at all but rather a propagation through a guided medium. For this case, due to propagation dispersion effects $t_p$ and $t_s$ are not identical, but close enough that throughputs in the range of $10^{10}$ bits/sec are possible. Assuming a group velocity of $10^8$ meters/sec, and a fiber length of 50 kilometers, the settling time is 0.5 millisec. At a conservative bit rate of 100 Megabits/sec, there are 50,000 bits propagating through the fiber at any time ("concurrent computations" in the language above). At this velocity and bit rate, the maximum distance for which there is no concurrency in the communication medium is one meter. □

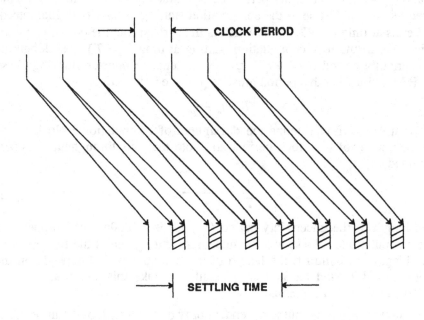

**Figure 6-3.** An example of a synchronous interconnect with a clock period much smaller than the reciprocal of the settling time, due to a small uncertainty period.

**Example 6-2.**
For typical practical Boolean logic circuits, designed to minimize the settling time rather than maximize the propagation time, $t_p$ is typically very small, and concurrent computations within the computational block are not possible. The maximum throughput is the reciprocal of the settling time. □

**Example 6-3.**
Consider a hypothetical (and perhaps impractical) circuit technology and logic design strategy which is designed to achieve $t_p \approx t_s$. In this case the throughput can be much higher than the reciprocal of the settling time, and many concurrent computations within the computational block are possible. □

While example 6-3 is not likely to be achieved, example 6-2 and example 6-3 suggest the possibility of designing circuits and logic to minimize the uncertainty period $(t_s - t_p)$ (even at the expense of increasing $t_s$) rather than minimizing $t_s$ as is conventional. For example, one could insure that every path from input to output had the same number of gates, and carefully match the gate settling times. In such a design style, the throughput could be increased to exceed the reciprocal of the settling time. This has recently been considered in the literature, and is called *wave pipelining* [4].

## 6.1.2. Pipelining

The form of concurrency associated with figure 6-3 is known as *pipelining*. A useful definition of pipelining is the ability to *initiate* a *new* computation at the input to a computational block prior to the *completion* of the *last* computation at the output of that block. Since this results in more than a single computation in process within the block at any given time, pipelining is a form of concurrency, always available when $t_p > 0$. The number of *pipeline stages* is defined as the number of concurrent computations in process at one time. For example, if we take the liberty of calling the fiber propagation in example 6-1 a "computation", then the fiber has 50,000 pipeline stages.

In conventional digital system design, $t_p$ for computational blocks is typically small, and pipelining requires the addition of *pipeline registers*. To see this potential, make the idealistic assumption that the computational block of figure 6-2 can be split into $N$ sub-blocks, the output of each connected to the input of the next, where each sub-block has a propagation time of $t_p/N$ and a settling time of $t_s/N$. If this is possible, then the block can be pipelined by inserting $(N-1)$ pipeline registers between each pair of these sub-blocks as shown in figure 6-4. Each of these registers, according to the analysis, can use a clock frequency of $N/(t_s - t_p)$ because the uncertainty period is correspondingly smaller, assuming that each clock phase is adjusted to fall within the certainty period relative to the last clock phase.

To see that the uncertainty period is reduced for figure 6-4, in figure 6-2 the middle of the certainty period is delayed relative to the clock time $t_0$ by $(t_p + t_s + T)/2$. Assuming that the clock phase for each pipeline register in figure 6-4 is chosen in the middle of this certainty period, then relative to the

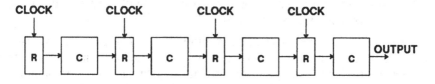

**Figure 6-4.** Pipelining of figure 6-2 for $N = 4$ sub-blocks and $(N - 1 = 3)$ intermediate pipeline registers.

previous clock phase the delay is $N$ times smaller, or $(t_p + t_s + T)/2N$. The total propagation and settling times for the pipeline are then

$$t_{p,\text{pipeline}} = (N-1)\cdot\frac{t_p + T + t_s}{2N} + \frac{t_p}{N} \qquad (6.3)$$

$$t_{s,\text{pipeline}} = (N-1)\cdot\frac{t_p + T + t_s}{2N} + \frac{t_s}{N} \qquad (6.4)$$

and the length of the uncertainty period is now

$$t_{s,\text{pipeline}} - t_{p,\text{pipeline}} = \frac{t_s - t_p}{N}, \qquad (6.5)$$

a factor of $N$ smaller. Thus, the theoretical maximum throughput is a factor of $N$ higher. Again, the reason for this increase is that the intermediate pipeline registers have reduced the length of the uncertainty period, since the pipeline registers have rendered the uncertainty period zero for all but the last block (by controlling the computational latency with the clock).

This interpretation of the role of pipeline registers as reducing the uncertainty period is unconventional. A more common approach is to *start* with the assumption that there are pipeline registers (and hence pipeline stages) and a given fixed clock frequency, and then place as much of the total computation within each stage as possible, with the constraint that the settling time has to be less than the clock period. In this common viewpoint, pipelining and pipeline registers are synonymous. Within their domain of common applicability, the uncertainty period and the common viewpoints are simply different ways of expressing the same design approach. However, the uncertainty period approach we have presented here is more general, in that it includes pipelining without pipeline registers, as often occurs in communication or interconnect. More importantly, this approach serves as a unified framework under which computational blocks, communication or interconnect, and combinations of the two can be characterized — where the computational blocks often utilize pipeline registers to reduce the uncertainty period, and communication links often do not (because the uncertainty period is inherently very small).

In some situations, the *total* settling time of the computation is just as important as the throughput, for example when the computation sits in a feedback loop. Thus, the effect of pipeline registers on this computational latency is also of interest. If we use the minimum clock period $T = t_s - t_p$ in (6.4), the total settling time through the pipeline is

$$t_{s,\text{pipeline}} = t_s, \qquad (6.6)$$

and the total settling time is the *same* before and after the insertion of the pipeline registers. Thus, we have not paid a penalty in total settling time in return for the increase in throughput by a factor of $N$, since only the variability in settling time has been reduced.

In practice, depending on the system constraints there are two interpretations of computational latency, as illustrated in the following examples.

**Example 6-4.**
In a computer or signal processing system, the pipeline registers introduce a *logical delay*, analogous to the $z^{-1}$ operator in Z-transforms. Expressed in terms of these logical delays, the $N$ pipeline registers increase computational latency by $N$ (equivalent to a $z^{-N}$ operator). This introduces difficulties, such as unused pipeline stages immediately following a jump instruction, or additional logical delays in a feedback loop. □

**Example 6-5.**
In some circumstances, the computational latency as measured in time is the critical factor. For example, in the media access controller for a local-area network, the time to respond to an external stimulus is critical. For this case, as we have seen, the addition of pipeline registers need not increase the computational latency at all. With or without pipeline registers, the computational latency is bounded below by the inherent precedences in the computation as implemented by a particular technology. □

In practice, it is usually not possible to precisely divide a computational block into "equal-sized" pieces. In that case, the throughput has to be adjusted to match the *largest* uncertainty period for a block in the pipeline, resulting in a lowered throughput. There are a number of other factors, such as register setup times, which reduce the throughput and increase the overall settling time relative to the fundamental bounds that have been discussed here. One of the most important of these is the effect of interconnect delay and clock skew, which we will address next.

## 6.1.3. Clock Skew in Synchronous Interconnect

The effects of clock phase and *interconnect delay* (delay of signals passing between computational blocks) will now be considered. Clearly, any uncertainty in clock phase will reduce the throughput relative to (6.2), since earlier results required precise control of clock phase within a vanishing certainty period. Conversely, any fixed delay in the interconnect will not necessarily affect the achievable throughput, because it will increase the propagation and settling times equally and thus not affect the length of the uncertainty period. In practice, for common digital system design approaches, the effect of any uncertainty in clock phase is magnified by any interconnect delays.

In this section, we relax the previous assumptions, and assume that the clock phase can be controlled only within some known range (similar to the uncertainty period for the computational block). We then determine the best throughput that can be obtained following an approach similar to [5,6].

We can analyze clock skew with the aid of figure 6-5, in which we modify figure 6-2(b) to introduce some new effects. In particular, we model the following:

- *Propagation and settling time.* As before, the propagation and settling time of the computation are $t_p$ and $t_s$, except that we now include in these times any latencies relative to the clock transition imposed by the implementation of R1.

- *Interconnect delay.* We assume interconnect delay $d$ due to the diffusion of the signal through the interconnect wires between R1 and C1 and between C1 and R2.

- *Artificially added delay.* We assume that another delay $\varepsilon$ is artificially added to the interconnect delay. The $\varepsilon$ delay could be introduced, for example, by making R1 a double register with two clock phases, thereby introducing an artificial delay. We will find that $\varepsilon$ is helpful in controlling the effects of clock skew, by effectively making the minimum interconnect delay larger than zero.

- *Clock skew.* We assume that the clock transition at R1 occurs at time $t_0$ at R1, and the clock phase at R2 is $t_0 + \delta$, where $\delta$ is the clock skew. This clock skew can be either inadvertent, due for example to processing variations or interconnect delays in the clock distribution, or it can be deliberately controlled, for example to adjust the R2 clock phase to fall within the certainty period. Further, it is possible for $\delta$ to be either

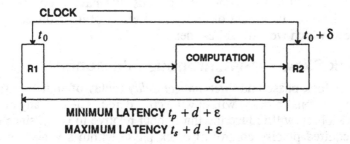

**Figure 6-5.** Illustration of the effects of clock skew $\delta$, where two computations C1 and C2 are synchronized using this clock.

positive or negative.

There is a certainty region of parameters $\{t_p, t_s, d, \varepsilon, \delta\}$ where reliable operation is assured in figure 6-5. This is analyzed in Appendix A for three cases:

- *Idealistic case.* If there is no uncertainty in $d$ or $\delta$ and we set the skew $\delta$ most advantageously, then the bound of (6.2) is achieved. As expected the interconnect delay need not necessarily slow the throughput.

- *Pessimistic case.* Assume that $\varepsilon$ can be precisely controlled (since it is controlled by relative clock phases) but that $\delta$ will inevitably only be controllable within a range, say $|\delta| < \delta_{max}$. Thus, we are not attempting to set $\delta$ most advantageously. Further, we assume that all that is known about the interconnect delay is that it is bounded, $|d| < d_{max}$. The throughput is then bounded by

$$\frac{1}{T} < \frac{1}{t_s - t_p + d_{max} + 2\delta_{max}} . \tag{6.7}$$

For a system with a complex interconnect pattern, it would be very difficult to control the relationship of $\delta$ and $d$. In this case, we should expect $\delta_{max} \approx d_{max}$, and the throughput would be bounded by

$$\frac{1}{T} < \frac{1}{(t_s - t_p) + 3d_{max}} . \tag{6.8}$$

- *Optimistic case.* For simple topologies like a one-dimensional pipeline, much higher throughput can be obtained by routing the signals and clocks in such a way that $d$ and $\delta$ can be coordinated with one another [5]. Assume that the interconnect delay is known to be $d_0$ with variation $\Delta d$, and the skew is chosen to be $\delta_0$ with variation $\Delta\delta$. Further assume that $\varepsilon$ and $\delta_0$ are chosen most advantageously to maximize the throughput. Then the throughput is bounded by

$$\frac{1}{T} < \frac{1}{(t_s - t_p) + 2(\Delta\delta + \Delta d)} , \tag{6.9}$$

a considerable improvement over (6.8) if the delay and skew variations are small.

This analysis shows that the reliable operation of the idealized synchronous interconnection of Section 6.1.1 can be extended to accommodate interconnect delays and clock skew, even with variations of these parameters, albeit with some necessary reduction in throughput.

To get a feeling for the numbers, consider a couple of numerical examples.

**Example 6-6.**

Consider the pessimistic case, which would be typical of a digital system with an irregular interconnection topology that prevents easy coordination of interconnect delay and clock skew. For a given clock speed or throughput, we can determine from (6.8) the largest interconnect delay $d_{max}$, that can be tolerated, namely $(T - t_s)/3$, assuming that the interconnect delay and clock skew are not coordinated and assuming the worst-case propagation delay, $t_p = 0$. For a 100 MHz clock frequency, a clock period is 10 nsec, and assuming the settling time is 80% of the clock period, the maximum interconnect delay is 667 psec. The delay of a data or clock signal on a printed-circuit board is on the order of five to ten picoseconds per millimeter (as compared to a free-space speed of light of 3.3 picoseconds per millimeter). The maximum interconnect distance is then 6.7 to 13.3 cm. Clearly, synchronous interconnect is not viable on PC boards at this clock frequency under these pessimistic assumptions. This also doesn't take into account the delay in passing through the pins of a package, roughly one to three nanoseconds (for ECL or CMOS respectively) due to capacitive and inductive loading effects. Thus, we can see that interconnect delays become a very serious limitation in the board-level interconnection with 100 MHz clocks. □

**Example 6-7.**

On a chip the interconnect delays are much greater (about 90 picoseconds per millimeter for $Al–SiO_2–Si$ interconnect), and are also somewhat variable due to dielectric and capacitive processing variations. Given the same 667 psec interconnect delay, the maximum interconnect distance is now about 8 mm. (This is optimistic since it neglects the delay due to source resistance and line capacitance — which will be dominant effects for relatively short interconnects.) Thus we see difficulties in using synchronous interconnect on a single chip for a complex and global interconnect topology. □

Again, it should be emphasized that greater interconnect distance is possible if the clock skew and interconnect delay can be coordinated, which may be possible if the interconnect topology is simple as in a one-dimensional pipeline. This statement applies at both the chip and board levels.

## 6.1.4. Parallel Signal Paths

An important practical importance of pipeline registers is in synchronizing the signals on parallel paths. The transition phase offset between these parallel paths tends to increase through computational blocks and interconnect, and can be reduced by a pipeline register to the order of the clock skew across the bits of this multi-bit register. Again, the register can be viewed as reducing the size of the uncertainty region, in this case spatially as well as temporally.

In section 6.1.1 we described the total uncertainty region for a collection of parallel signals as the union of the uncertainty regions for the individual signals. From the preceding, the total throughput is then bounded by the reciprocal of the length of this aggregate uncertainty period. In contrast, if each signal path from among the parallel paths were treated independently (say using the mesochronous techniques to be described later), the resulting throughput could in principle be increased to the reciprocal of the maximum of the individual uncertainty periods. For many practical cases, we would expect the longest uncertainty period to include the other uncertainty periods as subsets, in which case these two bounds on throughput would be equal; that is, there is no advantage in treating the signals independently. The exception to this rule would be where the uncertainty periods were largely non-overlapping due to a relative skew between the paths that is larger than the uncertainty period for each path, in which case there would be considerable advantage to dealing with the signals independently.

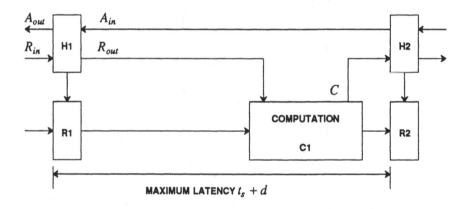

**Figure 6-6.** An anisochronous interconnect circuit, in which relative to figure 6-5 the clock has been removed and replaced by two handshake circuits H1 and H2. H1 and H2 generate the clocks for R1 and R2 respectively. The signals $R$ and $A$ are respectively the request and acknowledge handshaking signals, essentially a locally-generated clock, and $C$ is the completion signal for the logic block.

## 6.2. ANISOCHRONOUS INTERCONNECT

Having understood synchronous interconnect, it is appropriate to compare it against the self-timed anisochronous interconnect considered in earlier chapters.

The anisochronous interconnect is shown in figure 6-6. The clock in figure 6-5 has been replaced by a pair of handshake blocks H1 and H2, which generate the request signal $R_{out}$ for the next block and accept the acknowledge signal $A_{in}$ from that same block. In addition, the calculation block now generates a completion signal $C$, indicating that the setting time has been completed, and accepts an input signal $R_{out}$ which initiates its computation.

A *signal transition graph* for H1 is shown in figure 6-7 for a four-phase handshaking circuit appropriate in figure 6-6. This diagram models the order in which transitions in the H1 occur, and also the precedences that must be maintained by the circuitry in H1. For example, an arc from $A_{out}^{+}$ to $R_{out}^{+}$ indicates that the positive transition in $A_{out}$ must precede the positive transition in $R_{out}$.

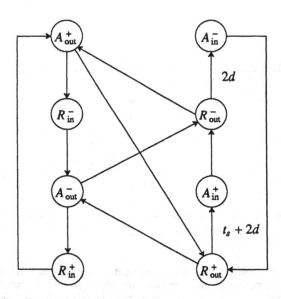

**Figure 6-7.** A signal transition graph for a four-phase pipeline handshaking circuit H1 from [7]. The superscript "+" or "-" indicates a positive or negative transition in the corresponding signal. Arcs are labeled with the corresponding latency.

The maximum throughput with which the system can operate is determined by the largest latency in any loop in figure 6-7. That maximum latency is the loop on the right, $R_{out}^+ \rightarrow A_{in}^+ \rightarrow R_{out}^- \rightarrow A_{in}^-$. If the settling time between the registers is the computational settling time $t_s$ plus the interconnect delay $d$ as shown, then the latency of the $R_{out}^+ \rightarrow A_{in}^+$ transition must be $t_s + 2d$ because this loop through H2 includes two interconnect delays plus the settling time, and the latency in the transition $R_{out}^- \rightarrow A_{in}^-$ must similarly be $2d$ because there is no computation in this loop. Additional latencies due to the handshake circuitry have been neglected. The total throughput is thus bounded by

$$\frac{1}{T} \leq \frac{1}{t_s + 4d}. \tag{6.10}$$

There are two-phase handshake approaches that are less reliable but reduce the $4d$ in the denominator to $2d$. Comparing (6.10) to (6.8), we see that the anisochronous interconnect throughput depends on the *actual* delay, whereas the pessimistic bound on throughput for synchronous interconnect depends on the *maximum* delay. (In the case of a pipeline, the total throughput will be dominated by the block in the pipeline with the maximum delay, so the two will be essentially the same.) On the other hand, the synchronous interconnect can take advantage of a non-zero propagation time $t_p$ to pipeline without pipeline registers, whereas the anisochronous interconnect does not. At its best, synchronous interconnect can operate with a throughput that is independent of interconnect delay, and only limited by delay *variation*, as demonstrated in (6.9). Thus, depending on the circumstances, either the synchronous or anisochronous interconnect can achieve the higher throughput. However, in the presence of large interconnect delays, the synchronous interconnect clearly has a potentially higher throughput.

An important point is that the throughputs of both the synchronous and anisochronous are generally adversely affected by the interconnect delay, and especially so for complicated interconnect topologies. As technologies scale, this restriction will become a more and more severe limitation on the performance of digital systems. However, as demonstrated by digital communication, which has experienced large interconnect delays from its inception, this delay-imposed throughput limitation is not fundamental, but is imposed in the synchronous case by the open-loop nature of the setting of clock phase.

An advantage of both synchronous and anisochronous interconnect is that they are free of metastability. This is avoided by insuring through the design methodology that the clock and data are never lined up precisely, making it difficult for clocked memory elements to reliably sample the signal.

## 6.3. SYNCHRONIZATION IN DIGITAL COMMUNICATION

In digital communication, the interconnect delays are very large, so that alternative synchronization techniques are required [3]. These approaches are all isochronous, implying that the signals are all slaved to clocks, but differ as to whether a common clock distributed to each node of the network is used (mesochronous) or independent clocks are used at the nodes (plesiochronous and heterochronous). They also share a common disadvantage relative to synchronous and anisochronous interconnect — the inevitability of metastable behavior. Thus, they all have to be designed carefully with metastability in mind, keeping the probability of that condition at an acceptable level.

Limitations to throughput due to propagation delays are avoided in digital communication as shown in figure 6-8. First, two communicating modules are each provided a clock, C1 and C2. A clock is also derived from the incoming signal in the *timing recovery circuit*, and is denoted C3. These clocks have the following relationships:

- C3 is synchronous with the signal at the input to the FIFO, since it is derived from that signal.

- C3 is mesochronous to C1, since it has the same average frequency as dictated by the common signal but has an indeterminate phase due to the interconnect delay. It can also have significant phase jitter due to a number of effects in the long-distance transmission [3].

- C1 and C2 are either mesochronous, if they originated from a common source, or they are independent. In the later case, they are either plesiochronous or heterochronous.

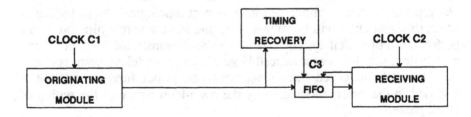

**Figure 6-8.** An illustration of how throughput is made independent of delay in digital communication.

The purpose of the *FIFO* (also called an *elastic store*) is to adjust for the differences in phase, likely to be time varying, between C3 and C2. For mesochronous C1 and C2, this phase difference is guaranteed to be bounded, so that the FIFO can be chosen with sufficient storage capacity to never overflow. For heterochronous C1 and C2, where the average frequency of C1 is guaranteed to be lower than the average frequency of C2, the FIFO occupancy will decrease with time, and hence no data will ever be lost. For plesiochronous C1 and C2, the FIFO could overflow, with the loss of data, if the average frequency of C1 happens to be higher than C2. This loss of data is acceptable on an occasional basis, but may not be permissible in a digital system design.

## 6.4. NON-SYNCHRONOUS INTERCONNECT

We found previously that the performance of both synchronous and anisochronous interconnects in digital systems are limited as a practical matter by the interconnect delays in the system. With the anisochronous approach, this limitation was fundamental to the use of round-trip handshaking to control synchronization. In the synchronous (but not anisochronous) case, we showed that this limitation is not fundamental, but rather comes from the inability to tightly control clock phases at synchronization points in the system. The reason is the "open-loop" nature of the clock distribution, making us susceptible to processing variations in delay. If we can more precisely control clock phase using a "closed-loop" approach, the throughput of the synchronous approach can more nearly approach the fundamental limit of (6.2), and considerably exceed that of anisochronous interconnect in the presence of significant interconnect delays. In this section, we explore some possibilities in that direction, borrowing techniques long used in digital communication.

### 6.4.1. Mesochronous Interconnect

Consider the case where a signal has passed over an interconnect and experienced interconnect delay. The interconnect delay does not increase the uncertainty period, and thus does not place a fundamental limitation on throughput. If this signal has a small uncertainty period, as for example it has been resynchronized by a register, then the certainty period is likely to be a significant portion of the clock cycle, and the phase with which this signal is resampled by another register is not even very critical. The key is to avoid a sampling phase within the small uncertainty period, which in synchronous interconnect can be insured only by reducing the throughput. But if the sampling phase can be controlled in closed-loop fashion, the interconnect delay should not be a factor, as demonstrated in digital communication systems.

Another perspective on clock skew is that it results in an indeterminate phase relationship between local clock and signal; in other words, the clock and signal are actually mesochronous. In *mesochronous interconnect,* we live with this indeterminate phase, rather than attempting to circumvent it by careful control of interconnect delays for clock and signal. This style of interconnect is illustrated in figure 6-9. Variations on this method were proposed some years ago [8] and pursued into actual chip realizations by a group at M.I.T. and BBN [9,10] (although they did not use the term "mesochronous" to describe their technique). We have adapted our version of this approach from the mesochronous approach used worldwide in digital communication, except that in this case we can make the simplifying assumption that the phase variation of any signal or clock arriving at a node can be ignored. The primary cause of the residual phase modulation will be variations in temperature of the wires, and this should occur at very slow rates

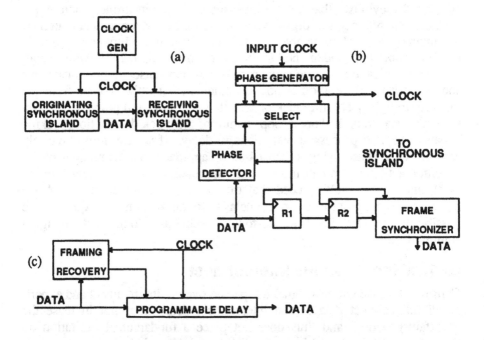

**Figure 6-9.** A mesochronous interconnection for a digital system. (a) Clock and signal distribution are identical to the synchronous case of figure 6-1a. Each functional unit corresponds to a synchronous island. (b) Interface circuit required for mesochronous-to-synchronous conversion. (c) Optional frame synchronizer circuit (within the synchronous island).

and at most requires infrequent phase adjustments. This simplification implies that clocks need not be derived from incoming signals, as in the timing recovery of figure 6-8, but rather a distributed clock can be used as a reference with which to sample the received signal. However, we must be prepared to adjust the *phase* of the clock used for sampling the incoming signal to account for indeterminate interconnect delays of both clock and signal. We thus arrive at figure 6-9.

First, we divide the digital system into functional entities called "synchronous islands". The granularity of partitioning into synchronous islands is guided by the principle that within an island the interconnect delays are small relative to logic speed, and traditional synchronous design can be used with minimal impact from interconnect delays. The maximum size of the synchronous island depends to a large extent on the interconnect topology, as we have seen previously. In near-term technology, a synchronous island might encompass a single chip within a digital system, and for the longer term a single clip may be partitioned into two or more synchronous islands. The interconnection of a pair of synchronous islands is shown in figure 6-9a; externally, the connection is identical to the synchronous interconnection in figure 6-1. The difference is in the relaxed assumptions on interconnect delays.

The mesochronous case requires a more complicated internal interface circuit, as illustrated in figure 6-9b. This circuit performs the function of *mesochronous-to-synchronous conversion*, similar in function but simpler than the FIFO in figure 6-8. This conversion requires:

- A *clock phase generator* to create a set of discrete phases of the clock. This does not require any circuitry running at speeds greater than the clock speed, but rather can be accomplished using a circuit such as a ring oscillator phase-locked to the external clock. A single phase generator can be shared amongst synchronous islands (such as one per chip), although to reduce the routing overhead it can be duplicated on a per-island or even on a multiple generator per-island basis.

- A *phase detector* determines the certainty period of the signal, for example by estimating the phase difference between a particular phase of the clock and the transitions of the signal. Generally one phase detector is required for each signal line or group of signal lines with a common source and destination and similarly routed.

- A *sampling register* R1 with sampling time chosen well within the certainty period of the signal, as controlled by the phase detector. This register reduces the uncertainty period of the signal, for example eliminating any finite rise-time effects due to the dispersion of the interconnect, and also controls the phase of the uncertainty period relative to the local clock. Depending on the effect of temperature variations in the system, this appropriate phase may be chosen once at power-up, or may be adjusted infrequently if the phase variations are a significant portion of a clock cycle.

- A second register R2 resamples the signal using the clock phase used internally to the synchronous island. At this point, the signal and this internal clock are synchronous. By inference, all the signals at the input to the synchronous island are also synchronous.

- If the system interconnect delays are larger than one clock cycle, it may be convenient at the architectural level to add an optional *frame synchronizer*, which is detailed in figure 6-9c. The purpose of this block is to synchronize all incoming signal lines at the word-level (for example at byte boundaries) by adding a programmable delay. This framing also generally requires additional framing information added to the signal.

The practicality of the mesochronous interconnection depends to a large extent on the complexity of the interconnection circuit and the speed at which it can be implemented. All the elements of this circuit are simple and pretty standard in digital systems with the possible exception of the phase detector. Thus, the design of mesochronous interconnection hinges largely on innovative approaches to implementing a phase detector. A phase detector can be shared over a group of signals with the same origin and destination and same path length, and it can even be time-shared over multiple such functions since we don't expect the phase to change rapidly. Nevertheless, the phase detector should be realized in a small area, at least competitive with anisochronous handshaking circuits. Further, it should be fast, not restricting the clock rate of the interconnect, since the whole point is to obtain the maximum speed. This should not be a problem on slower out-of-package interconnections, but may be difficult to achieve on-chip. Any phase detector design is also likely to display metastable properties, which must be minimized. The design of the phase detector may also place constraints on the signal, such as requiring a minimum number of transitions. In view of all these factors, the phase detector is a challenging analog circuit design problem.

## 6.4.2. Heterochronous and Plesiochronous Interconnect

Like synchronous interconnect, and unlike anisochronous interconnect, mesochronous interconnect requires distribution of a clock to all synchronous islands, although any interconnect delays in this distribution are not critical. If the power consumption of this clock is of concern, in principle it would be possible to distribute a sub-multiple of the clock frequency, and phase-lock to it at each synchronous island. But if the interconnect wires for clock distribution are of concern, they can be eliminated as in the digital communication system of figure 6-8, where the timing recovery is now required. There are two useful cases:

- If clocks C1 and C2 are *heterochronous*, such that C1 is guaranteed to have a lower frequency than C2, then the FIFO can be very small and overflow can never occur. However, the signal from the FIFO to the receiving module would have to contain "dummy bits", inserted whenever necessary to prevent FIFO underflow, and a protocol to signal where those bits occur.

- If clocks C1 and C2 are *plesiochronous*, then overflow of the FIFO can occur in the absence of flow control. Flow control would insert the "dummy bits" at the originating module whenever necessary to prevent FIFO overflow. Flow control could be implemented with a reverse handshaking signal that signaled the originating module, or if another interconnect link existed in the opposite direction, that could be used for this purpose.

## 6.5. ARCHITECTURAL ISSUES

While synchronous interconnect is still the most common, both anisochronous and mesochronous interconnects are more successful in abstracting the effects of interconnect delay and clock skew. Specifically, both design styles insure reliable operation independent of interconnect delay, without the global constraints of sensitivity to clock distribution phase. Each style of interconnect has disadvantages. Mesochronous interconnect will have metastability problems, and requires phase detectors, while anisochronous interconnect requires extra handshake wires, handshake circuits, and a significant silicon area for completion-signal generation.

The style of interconnection will substantially influence the associated processor architecture. The converse is also true — the architecture can be tailored to the interconnect style. As an example, if at the architectural level we can make a significant interconnect delay one stage of a pipeline, then the effects of this delay are substantially mitigated. (Consider for example (6.10) with $t_s = 0$.)

Many of these issues have been discussed with respect to anisochronous interconnection in [11]. On the surface one might presume that mesochronous interconnection architectural issues are similar to synchronous interconnection, which if true would be a considerable advantage because of the long history and experience with synchronous design. However, the indeterminate delay in the interconnection (measured in bit intervals at the synchronous output of the mesochronous-to-synchronous conversion circuit) must be dealt with at the architectural level. For synchronous interconnection, normally the delay between modules is guaranteed to be less than one clock period (that doesn't have to be the case as illustrated in figure 6-3), but with mesochronous interconnection the delay can be multiples of a bit period. This has fundamental implications to the architecture. In particular, it implies a higher level of synchronization (usually called "framing" in digital communications) which line up signal word boundaries at computational and arithmetic units.

In the course of addressing this issue, the following key question must probably be answered:

- Is there a maximum propagation delay that can be assumed between synchronous islands? If so, is this delay modest?

The answer to this question is most certainly yes in a given chip design, but difficult to answer for a chip designed to be incorporated into a mesochronous board- or larger-level system. As an example of an architectural approach suitable for worst-case propagation delay assumptions, we can include frame synchronizers like figure 6-9c which "build-out" each interconnection to some worst-case delay. Every interconnection thus becomes worst-case, and more importantly predictable in the design of the remainder of the architecture. However, this approach is probably undesirable for reasons that will soon be elaborated. Another approach is to build into the architecture adjustment for the actual propagation delays, which can easily be detected at power-up. Yet another approach is to use techniques similar to packet switching in which each interconnection carries associated synchronization information (beginning and end of message, etc.), and design the architecture to use this information. We have studied this problem in some detail in the context of inter-processor communication [12].

As previously mentioned, each data transfer in a anisochronous interconnect requires two to four propagation delays (four in the case of the most reliable four-phase handshake). In contrast, the mesochronous interconnection does not have any feedback signals, and is thus able to achieve whatever throughput can be achieved by the circuitry, independent of propagation delay. This logic applies to feedforward-only communications. The more interesting case is the command-response situation, or more generally systems with feedback, since delay will have a considerable impact on the performance of such systems. To some extent the effect of delay is fundamental and independent of interconnect style: the command-response cycle time cannot be smaller than the round-trip propagation delay. However, using the "delay build-out" frame synchronizer approach described earlier would have the undesirable effect of unnecessarily increasing the command-response time of many interconnections in the system. Since this issue is an interesting point of contrast between the anisochronous and mesochronous approaches, we will now discuss it in more detail.

## 6.5.1. Command-Response Processing

Suppose two synchronous islands are interconnected in a bilateral fashion, where one requests data and the other responds. An example of this situation would be a processor requesting data from a memory — the request is the address and the response is the data residing at that address. Assuming for the moment that there is no delay generating the response within the responding synchronous island, the command-response cycle can be modeled simply as a delay by $N$ clock cycles, corresponding to roughly a delay of $2t_p$, as illustrated in figure 6-10.

It is interesting that this delay is precisely analogous to the delay (measured in clock cycles) introduced by pipelining — we can consider this command-response delay as being generated by $N$ pipeline registers. As in pipelining, this delay can be deleterious in reducing the throughput of the processing, since the result of an action is not available for $N$ clock cycles. One way of handling this delay is for the requesting synchronous island to go into a wait state for $N$ clock cycles after each request. The analogous approach in pipelining is to launch a new data sample each $N$ cycles, which is known as $N$-slow [13], and the performance will be inversely proportional to $N$. This approach is analogous to the anisochronous interconnection (which may require considerably more delay, such as four propagation delays for the transfer in each direction).

For the mesochronous case, there are some architectural alternatives that can result in considerably improved performance under some circumstances, but only if this issue is addressed at the architectural level. Some examples of these include:

- If we have some forward-only communications coincident with the command-response communications, we can *interleave* these feedforward communications on the same lines.

REQUEST ← $\boxed{\begin{array}{c} N \\ \text{CYCLE} \\ \text{DELAY} \end{array}}$ → RESPONSE

**Figure 6-10.** A model for command-response in a mesochronous interconnection. The delay is with respect to cycles of the common clock.

- If we have a set of $N$ independent command-response communications, we can interleave them on the interconnection. This is analogous to *pipeline interleaving* [14] (which can make full use of the throughput capabilities of a pipelined processing element provided that we can decompose the processing into $N$ independent streams). If the responding island cannot accommodate this high a throughput, then it can be duplicated as many times as necessary (in pipelining an analogous example would be memory interleaving).

- If we cannot fully utilize the interconnection because of the propagation delay, then at least we should be able to allow each processing element to do useful processing (at its full throughput) while awaiting the response. This is analogous to what is sometimes done when a synchronous processor interacts asynchronously with a slower memory.

The last of these options would be available to a anisochronous system, since the communication portion could be made a separate pipeline stage. However, the first two options, two forms of interleaving of communications, are not available in anisochronous systems because the *total* throughput is bounded by the propagation delay. If the communication bottlenecks are taken into account at the architectural level, it appears that mesochronous interconnection offers considerable opportunity for improved performance.

## 6.6. CONCLUSIONS

In this chapter we have attempted to place the comparison of digital system synchronization on a firm theoretical foundation, and compare the fundamental limitations of the synchronous, anisochronous, and mesochronous approaches. A firm conclusion is that interconnect delays place a fundamental limitation on the communication throughput for anisochronous interconnect (equal to the reciprocal of two or four delays), and this limitation does not exist for isochronous interconnect. Further, isochronous interconnect can actually achieve pipelining in the interconnect (as illustrated in example 6-1) without additional pipeline registers, whereas anisochronous cannot. Further, anisochronous requires extra interconnect wires and completion signal generation. Thus, as clock speeds increase and interconnect delays become more important, isochronous interconnect shows a great deal of promise. However, the advantages of any synchronization technique cannot be fully exploited without modifications at the architectural level.

# APPENDIX 6-A
# CERTAINTY REGION FOR CLOCK SKEW

In this Appendix, we determine the certainty region for the parameters in figure 6-5. Consider a computation initiated by the clock at R1 at time $t_0$. Extending our earlier results, the conditions for reliable operation are now as follows:

- The earliest the signal can change at R2 is after the clock transition $t_0 + \delta$, or

$$t_0 + \delta < t_0 + t_p + d + \varepsilon. \tag{6.11}$$

- The latest time the signal has settled at R2 must be before the next clock transition $t_0 + \delta + T$, where $T$ is the clock period,

$$t_0 + t_s + d + \varepsilon < t_0 + \delta + T. \tag{6.12}$$

Simplified, these equations become

$$\delta < t_p + d + \varepsilon \tag{6.13}$$

$$T + \delta > t_s + d + \varepsilon. \tag{6.14}$$

Together, (6.13) and (6.14) specify a *certainty region* for $\{\delta, T\}$ where reliable operation is guaranteed.

With the aid of figure 6-11, we gain some interesting insights. First, if the skew is sufficiently positive, reliable operation cannot be guaranteed because the signal at R2 might start to change before the clock transition. If the skew is negative, reliable operation is always guaranteed if the clock period is large enough.

*Idealistic case.* The most advantageous choice for the skew is $\delta = t_p + d + \varepsilon$ at which point we get reliable operation for $T > t_s - t_p$. Choosing this skew, the fundamental limit of (6.2) would be achieved. This requires precise knowledge of the interconnect delay $d$ as well.

*Pessimistic case.* In figure 6-11 we see the beneficial effect of $\varepsilon$, because if it is large enough, reliable operation is guaranteed for any $\delta_{max}$. In particular, the condition is $t_p + d + \varepsilon > \delta_{max}$, or

$$t_p + d > \delta_{max} - \varepsilon. \tag{6.15}$$

Since the interconnect delay $d$ is always positive, (6.15) is guaranteed for

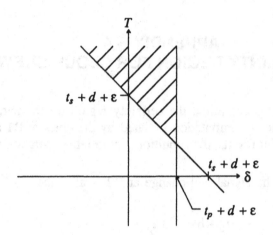

**Figure 6-11.** The crosshatched area is the certainty region, now in terms of $\{\delta,T\}$. Note that it is not always possible to choose the clock period $T$ sufficiently large to fall in the certainty region, but it is always possible to choose $T$ and $\varepsilon$ together large enough to accommodate any bounded clock skew.

any $t_p$ and $d$ so long as $\varepsilon > \delta_{max} - t_p$. Since $\varepsilon$ also has the effect of increasing $T$, it is advantageous to choose $\varepsilon$ as small as possible; namely, $\varepsilon = (\delta_{max} - t_p)$. Referring back to figure 6-11, reliable operation is guaranteed if

$$T > (t_s - t_p) + d + 2\delta_{max} . \tag{6.11}$$

Relative to the fundamental bound on clock period $(t_s - t_p)$, the clock period must be increased by $(d + 2\delta_{max})$. Taking account of the worst case $d = d_{max}$, we get (6.7).

*Optimistic case.* The throughput of (6.9) follows easily by the same method.

# REFERENCES

1. D. G. Messerschmitt, "Digital Communication in VLSI Design," *Proc. Twenty-Third Asilomar Conference on Signals, Systems, Computers*, (Oct. 1989).

2. D. G. Messerschmitt, "Synchronization in Digital Systems Design," *IEEE Trans. on Special Areas in Communications* **JSAC-8**(10)(October 1990).

3. E. A. Lee and D. G. Messerschmitt, "Digital Communication," *Kluwer Academic Press*, (1988).

4. D. Wong, G. De Micheli, and M. Flynn, "Designing High-Performance Digital Circuits Using Wave Pipelining," *IEEE ICCAD-89 Digest of Technical Papers*, (November 1989).

5. M. Hatamian and G. L. Cash, "Parallel Bit-Level Pipelined VLSI Designs for High-Speed Signal Processing," *IEEE Proceedings* **75**(9) p. 1192 (Sept. 1987).

6. M. Hatamian, "Understanding Clock Skew in Synchronous Systems," in *Concurrent Computations*, ed. S.C. Schwartz,Plenum (1988).

7. T. H. Meng, R. W. Brodersen, and D. G. Messerschmitt, "Automatic Synthesis of Asynchronous Circuits from High-Level Specifications," *IEEE Transactions on Computer Aided Design.* **8**(11)(November 1989).

8. W. Nix, *A System for Synchronous Data Transmission Over A Line of Arbitrary Delay*, M.S. Project Report, U.C. Berkeley (1981.).

9. P. Bassett, L. Glasser, and R. Rettberg, "Dynamic Delay Adjustment: A Technique for High-Speed Asynchronous Communication," *MIT Press*, (1986).

10. P. Bassett, *A High-Speed Asynchronous Communication Technique for MOS VLSI Systems*, Massachusetts Inst. of Technology (December 1985).

11. T. H.-Y. Meng, R. W. Brodersen, and D. G. Messerschmitt, "Asynchronous Design for Programmable Digital Signal Processors," *To be published in Trans. on ASSP*, (April 1991).

12. M. Ilovich, "High Performance Programmable DSP Architectures," *Ph.D. Thesis, University of California*, (April 1988).

13. K. Parhi and D.G. Messerschmitt, "Pipeline Interleaving and Parallelism in Recursive Digital Filters, Part I: Pipelining Using Scattered Look-Ahead and Decomposition," *IEEE Trans. on ASSP*, (July 1989).

14. E.A. Lee and D.G. Messerschmitt, "A Coupled Hardware and Software Architecture for Programmable Digital Signal Processors Part I: Hardware," *IEEE Trans. on Acoustics, Speech, and Signal Processing*, (September 1987 ).

# 7

---

# AUTOMATIC VERIFICATION

---

Steve M. Nowick and David L. Dill
Computer Systems Laboratory
Stanford University

The quest for faster computers will inevitably increase the use of concurrency in hardware designs. To some extent, modern processors, particularly multiprocessors, are best regarded as a collection of communicating processes (e.g. controllers, switching networks, and other processes). Unfortunately, it has long been recognized in the study of operating systems that concurrent systems are susceptible to subtle synchronization errors. We believe that this is already a problem in hardware design — one that is bound to increase with the use of concurrency.

Established approaches to CAD cannot solve the problem: simulation has proven to be inadequate for the task, because of the difficulty of simulating a large number of possible interleavings of parallel computations. Correct state-machine synthesis cannot prevent bugs that occur from the interactions among several state machines. Synthesis from a high-level concurrent program might help in some cases, but the problem of how to divide a large computation into cooperating controllers is probably too difficult to automate. Users will want to change their high-level programs to allow the synthesizer to generate alternative designs, but they will have no assurance that their changes preserve the correctness of the system.

These arguments militate for a new class of CAD tools: finite-state verifiers. These are programs that can compare state-machine representations of the behaviors of different processes, where the process can be described in various ways (for example, as a program or as a sequential circuit).

To clear up possible terminological confusion, when we say "verify," we mean to *prove* that a system meets some specification. We specifically do not mean testing (unless it is exhaustive) or any other method that ensures that the system is "probably" correct.

## 7.1. VERIFICATION OF SELF-TIMED CIRCUITS

There is one type of design style in which the problems of concurrency arise immediately at the implementation level: asynchronous. Asynchronous self-timed (unclocked) designs are of increasing interest because of the cost of broadcasting clocks over large areas of a chip. While a synchronous circuit can often be regarded as a single process, with all of the components executing in lock-step, a self-timed circuit is best thought of as a concurrent system — each component is a separate process that communicates with its neighbors by signalling on wires. Such a design must work in spite of timing variations among the components.

Of particular interest are speed-independent synchronization circuits, which are self-timed control circuits that function correctly assuming arbitrary delays in the components but no delays in the wires.

There is an existing automatic verifier for speed-independent self-timed circuits, which has been described elsewhere [1,2,3]. Previous work based on the verifier has concentrated on issues of specification, modeling, and the details of examples, not on detailed discussion of performance.

Several issues arise when evaluating a formal verification method:

1.    How much human effort is required? Theorem-proving methods, in particular, often require the user to provide hints about appropriate intermediate steps in the proof. It is usually difficult to get a handle on exactly how much effort is involved in a particular proof, and how much the use of a computer actually aids verification. *Our method requires human labor to model the components, write a structural description of the implementation, and describe the desired behavior. The actual verification is completely automatic.*

2.  What is the underlying model of system operation? Is the implementation described as a collection of analog transistors, transistor switches, gates, "elements" such as flip-flops, registers, or a program? The level of description may determine whether verification is feasible and what type of problems can be detected. *Our method models signals on individual wires (it cannot reasonably model data values). It can model elements with internal states. It is best suited for speed-independent self-timed circuits.*

3.  What specification is checked? It is almost a cliche that correctness of a circuit is meaningless without a specification. Verification methods vary in the types of properties that can be specified. Sometimes, it will be claimed that a system has been "verified" with no mention of what the specification actually was. Such claims are dangerous, because many serious problems arise because of a lack of common understanding of specifications. *Our method allows any finite-state specification. It cannot handle numerical times and it cannot handle "liveness" properties, such as progress and unbounded fairness.*

4.  How much computation is required? *Our method is exponential in the size of the specifications and the number of components of an implementation, in the worst case. However, it handles many practical examples in a few seconds.*

This work is still in the exploratory stages. Our verifier is written in Common Lisp, because it is more important to be able to modify our algorithms and data-structures quickly than to get incremental speedups by re-coding in a more efficient programming language. The examples here were run in Ibuki Common Lisp on a DEC VAXStation 3100.

We estimate that simple re-coding in C would yield about a factor of ten increase in speed, and using a faster computer would yield proportional speed-ups. Nevertheless, the program is the most effective we know of for its application.

In the remainder of this chapter, we describe the theory behind the verifier, and how it is used. We then specify three example circuits, all of them asynchronous arbiters: the first is a "tree arbiter" that has been used as a verification example previously[4,5]. The second and third examples are two approaches to verifying a new circuit: an arbiter that rejects a user if the resource is in use, allowing it to proceed. The first method compares the implementation and specification in the obvious way, and the second takes advantage of the hierarchical structure of the design. Performance figures for verification are given.

## 7.2. TRACE THEORY OF FINITE AUTOMATA

Trace theory is a formalism for modeling, specifying, and verifying speed-independent circuits; it is based on the idea that the behavior of a circuit can be completely described by a regular set of *traces*, or sequences of transitions. A single trace corresponds to a partial history of signals that might be observed at the input and output wires of the circuit.

The idea of using trace models of concurrent systems appears to originate with Hoare, who gave a trace semantics for a subset of CSP [6]; the specific application to self-timed circuits was first proposed by Rem, Snepscheut, and Udding [7]. The substantially modified version used here is by Dill [1,2].

This section provides a brief summary of trace theory based on finite automata (we assume familiarity with regular languages and finite automata [8]).

### 7.2.1. Trace Structures

A *trace structure* is a triple $T = (I, O, M)$ where $I$ is a finite set of *input wire names*, $O$ is a finite set of *output wires*, and $M$ is a deterministic finite automaton (DFA). $I$ and $O$ must be disjoint. We use the abbreviation $A$ for $I \cup O$. The automaton is a quadruple $(Q, \mathbf{n}, q_0, Q_F)$, where $Q$ is a finite set of *states*, $\mathbf{n}: Q \times A \rightarrow Q$ (which can be a partial function) is the *next-state function*, $q_0 \in Q$ is the *start states*, and $Q_F \subseteq Q$ is the set of *final states*. In a trace structure, $Q_F$ is always the same as $Q$; however, some of the automata produced by intermediate steps of the constructions below have $Q_F \neq Q$.

The regular language accepted by $M$ is the set of *successful traces*, written $S$, which are the partial histories the circuit can exhibit when "properly used"[1]. If it is improperly used, the results are not specified. In this way, a trace structure is like a warranty that can be voided if certain conditions are not met. An example of a proper-use constraint is the requirement that the *set* and *reset* inputs of a latch never be asserted simultaneously.

Consider a trace description of a non-inverting buffer (figure 7-1). The trace description depends on the initial conditions of the circuit; in this example, the logical values on the input and output wires are initially the same. The buffer waits until an input transition $a$ occurs, then produces an output $b$,

---

[1] Note that the case where $S = \varnothing$ cannot be represented in the automaton as defined above, for trivial reasons. This can easily be dealt with as a special case, so we will not discuss it further.

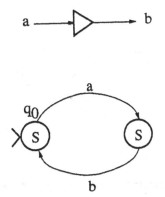

**Figure 7-1.** Non-inverting buffer and trace automata.

and so on. The undefined "next states" in the state diagram are as interesting as the defined ones: in the initial state there is no successor for $b$ because it is *impossible* for a $b$ transition to occur — $b$ is an output, so it is under the control of the circuit. The missing successor on $a$ in the next state models a *possible but undesirable* occurrence. It is possible because the input to a gate may change at any time (the gate cannot prevent it from happening). It is undesirable because two consecutive input changes may cause unpredictable output behavior — the output may pulse to the opposite value for an arbitrarily short time, or it may not change (because the input pulse was too short), or it may change "halfway" and then return to its original value. Such behavior in a speed-independent circuit is unacceptable, so our description of a buffer forbids inputs that can cause it.

The possible but undesirable traces are called *failures*. The set of all failures associated with a trace structure $T$ is written $F$. An automaton accepting $F$ can be derived from $M$ by adding a failure state $q_F$, making $q_F$ the sole accepting state (so $Q_F = \{q_F\}$), and adding successors according to the following rules:

- For every *input wire* $a$ and state $q$, if $n(q,a)$ is undefined, add a transition from $q$ on $a$ to $q_F$.

- Add a transition from $q_F$ to itself on every input and output.

The first rule means that if there is no transition on an input symbol from a state, the circuit is not "ready" for that input — a failure will result if it occurs. The second rule says that a failure cannot be redeemed by the arrival of additional inputs and that the subsequent behavior is completely unpredictable — any output may occur at any time. The automaton in figure 7-2 accepts the failure set of a buffer, where the final state is $q_F$ (any

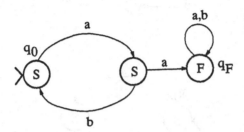

**Figure 7-2.** Trace automaton with failure state for non-inverting buffer.

string reaching a state marked $S$ is a success; any reaching $F$ is a failure). The failures are the traces that include the substring $aa$.

## 7.2.2. Operations on Trace Structures

There are three fundamental operations on trace structures: **hide** makes some output wires unobservable, **compose** finds the concurrent behavior of two circuits that have had some wires connected, and **rename** changes the names of the wires.

**hide** takes two arguments: a set of wires to delete and a trace structure. The deleted wires must be outputs, otherwise the operation is undefined. An automaton for the result of **hide** can be obtained by first adding a failure state (as above), then mapping each of the hidden wire names to $\varepsilon$, the empty string, to give a nondeterministic finite automaton with $\varepsilon$ transitions. This automaton can be converted to a deterministic automaton by using the subset construction with $\varepsilon$-closure (see [8]). Every state in the resulting automaton whose associated subset contains $q_F$ should then be deleted, giving the desired automaton.

In figure 7-3 shows, step-by-step, the effects of hiding $b$ in the buffer example of figure 7-2. This example illustrates the justification for the last step. Consider the trace $aa$ when $b$ is hidden. This could reflect several possible traces in the original circuit. It could be the $a$ transitions of the trace $aba$ — a success. Or it could be the $a$ transitions of $aa$ — a failure in the original circuit. If the inputs $aa$ are sent to the circuit without regard to $b$, the choice between success or failure is nondeterministic. For the purpose of implementing a specification, a circuit that "might fail" when given two $a$ inputs is as bad as a circuit that "always fails" given those inputs. In the construction of the previous paragraph, the states representing "possible failure" are those whose subsets contain $q_F$. Deleting these states has the desired effect of classifying as failures the traces that reach them.

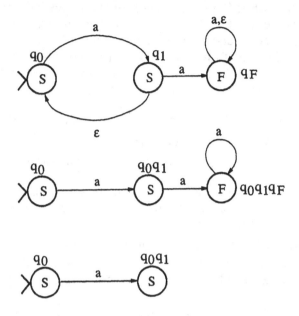

**Figure 7-3.** Hiding the output of a non-inverting buffer.

In this example, the successes are exactly $\varepsilon$ and $a$. Intuitively, there is no safe way to send more than one $a$ to a buffer unless the output can be observed to ensure that a $b$ occurs between each pair of $a$'s. (One can reasonably conclude from this example that a non-inverting buffer with hidden output is quite useless.)

**hide** is an important operation because it automatically suppresses irrelevant details of the circuit's operation — the unobservable signals on internal wires. In practice, hiding internal signals can result in a smaller model of behavior. It is also important because it allows trace structures to be compared based on their interface signals while ignoring internal signals. The ability to hide is one of the advantages of trace theory that is missing from earlier efforts [5].

**compose** models the effect of connecting identically-named wires between two circuits (called the *components*). Two circuits can be composed whenever they have no output wires in common. The composition of $T$ and $T'$ can be written $T \parallel T'$. If $T'' = T \parallel T'$, the set of outputs of $T''$ is $O'' = O \cup O'$ (whenever an output is connected to an input, the result is an output) and the set of inputs is $I'' = (I \cup I') - O''$ (an input connected to an input is an input). Note that $A'' = A \cup A'$.

The construction of $M''$, the automaton of the composite trace structure, consists of two steps. The first step is to define an automaton $\hat{M}$ using a modified product construction. First, add the failure state $q_F$ to $M$ and $q_F$ to $M'$, as in the definition of **hide**. The states in $\hat{M}$ are pairs of states in $M$ and $M'$: $\hat{Q} = (Q \times Q')$ and the start state, $\hat{q_0}$, is $(q_0, q_0')$. The definition of the successor function involves several cases:

$$\hat{n}(q_i, q_j') = \begin{cases} (n(q_i, a), q_j') & \text{if } a \in A - A' \\ (q_i, n'(q_j', a)) & \text{if } a \in A' - A \\ (n(q_i, a), n'(q_j', a)) & \text{otherwise} \end{cases} \qquad (7.1)$$

In effect, each component ignores wires that are not its inputs or outputs. If either of $n$ or $n'$ is used in the definition of the relevant case and is undefined, $\hat{n}(q_i, q_j')$ is undefined, also.

The second step is to delete every state that can reach a state of the form $(q_i, q_F')$ or $(q_F, q_j')$ by a string of zero or more outputs (the rationale for this is explained in the example below). This gives the final automaton.

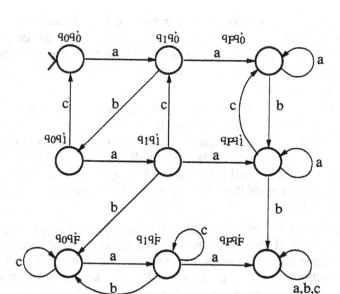

**Figure 7-4.** Serial composition of non-inverting buffers: First step.

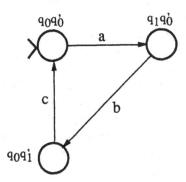

**Figure 7-5.** Serial composition of non-inverting buffers: Final result.

Figure 7-4 shows the initial step in the composition of two noninverting buffers in series, and figure 7-5 shows the final result. The automaton for the first buffer is as in figure 7-2; the automaton for the second buffer is the same, except that $b$ and $c$ are substituted for $a$ and $b$. To justify the last step in the construction, consider the state $(q_1, q_1')$, which is a success as shown in figure 7-4. However, whenever this state is entered there is a possibility that the first component will output a $b$, causing a failure in the second component. For the user of this circuit to guarantee that no failures can occur, $(q_1, q_1')$ must be avoided — in effect, it should be classified directly as a failure. Intuitively, this step exports an internal failure to the interface of the circuit.

**rename** takes as arguments a trace structure and a renaming function which maps its wire names to new names. The function must be a one-to-one correspondence — it is illegal to rename two different wires to be the same. The effect of **rename** on an automaton is simply to substitute the new names for the old names. This operation is useful for creating an instance of a circuit from a prototype.

## 7.2.3. Verification with Trace Theory

A trace structure specification of a circuit can be compared with a trace structure description of the actual behavior of the circuit, just as a logical specification can be compared with a program implementing it. When $T_I$ implements $T_S$, we say that $T_I$ *conforms to* $T_S$ (the inputs and outputs of the two trace structures must also be the same). This relation holds when $T_I$ can be *safely substituted* for $T_S$. More precisely, $T_I$ conforms to $T_S$ if, for every $T'$, whenever $T_S \parallel T'$ has no failures, $T_I \parallel T'$ has no failures, either. Intuitively, $T_I$ must be able to handle every input that $T_S$ can handle (otherwise, $T_I$ could fail in a context in which $T_S$ would have succeeded), and must not produce an output unless $T_S$ produces it (otherwise, $T_I$ could cause a failure in the surrounding circuitry when $T_S$ would not).

This relation can be tested by exchanging the input and output sets of $T_S$ to obtain $T_S^M$ (the *mirror* of $T_S$), in which $O_S^M = I_S$ and $I_S^M = O_S$, then composing $T_I \parallel T_S^M$ and checking whether the failure set of the composite trace structure is empty. This result is proved and justified in [2]. The intuition behind this result is that $T_S^M$ represents a context that will "break" any trace structure that is not a true implementation of $T_S$. Specifically, $T_S^M$ produces as an output everything that $T_S$ accepts as an input, so if $T_I$ fails on any of these, there will be a failure in $T_I \parallel T_S^M$. Similarly, $T_S^M$ accepts as input only what $T_S$ produces as output, so if $T_I$ produces anything else, there will be a failure in $T_I \parallel T_S^M$ also.

Conformation is a *partial order* rather than an *equivalence*, since an implementation is usually a *refinement* of a specification — there may be details of an implementation's operation that are not required by the specification. For this reason, conformation implies that an implementation meets *or exceeds* a specification.

Trace theory has the great advantage of supporting *hierarchical verification*. Using trace theory, a hierarchical design can be verified in the same way it is organized: a trace structure specification that has been verified at one level can be used in the description of the implementation of the next higher level. Hierarchical verification can be immensely more efficient than the alternative of "flattening" a hierarchical design. Suppose that a module $M$ consisting of several components is used in a larger design. One way to simplify verification is to write a specification $S$ that expresses only the relevant properties of $M$, verifying that the implementation of $M$ meets the specification $S$, then verifying the larger circuit using $S$ instead of the implementation of $M$. This suppresses irrelevant implementation details that otherwise would make verification of the larger circuit more difficult. Also, $M$ may be repeated many times in the larger design, but it only needs to be verified once using the hierarchical method.

## 7.2.4. An Automatic Verifier

There is an implementation of trace theory in Common Lisp that allows individual trace structures to be defined in a variety of ways. **hide**, **compose**, and **rename** have been implemented as Lisp functions, as has the conformation relation (the **conforms-to-p** function).

The program can be used as an automatic verifier. The user calls the verifier with a description of the implementation (consisting of trace structures for the primitives and an expression describing the topology of the circuit using **compose** and **hide** operations) and a specification (a trace structure for the

desired behavior). The **conforms-to-p** function then checks for conformation by composing the mirror of the specification with the implementation
and searching for failures in the resulting state graph. If the implementation
does not conform to the specification the verifier prints a failure trace to help
diagnose the problem.

For space and time efficiency, the state graph construction is search-driven:
states are generated only when needed. This simple implementation trick
often saves a tremendous amount of computation when the implementation
does not conform, because the program stops when it discovers the *first*
failure. This is particularly important since buggy circuits usually generate
many states representing random activity after the first problem occurs.

The program can translate several different user notations into trace structures (the structure of the program is such that it is easy to add pre-
processors). One such notation is a special-purpose macro for describing
Boolean functions: the user gives the input wires, a single output wire, a
Boolean formula, and the initial logical values of the wires. The program
translates this into a trace structure representing a gate that implements logic
function, with a single delay on the output (it is assumed that there are no
internal hazards).

The trace structure for a gate also includes an important environmental constraint. When the Boolean function of the inputs disagrees with the output
of the gate, we say the gate is *unstable* (it is *stable* otherwise). When the
gate is unstable, the environment is never allowed to then change the inputs
so that it becomes stable before the output changes. Otherwise, the output
of the gate could produce an arbitrarily short pulse (a hazard), which would
cause serious problems in most self-timed interconnection circuits.

Petri nets, which we will summarize briefly, can be used to describe more
general sequential behavior [9]. They are often used to describe speed-
independent circuits [10,11,12,13,14]. A Petri net consists of finite sets of
*places*, *bars*, and and *arcs*. An arc always connects a place $p$ to a bar $b$ or a
bar $b$ to a place $p$; in the first case, $p$ is called an *input place of b*, and in
the second, $p$ is called an *output place of b*. In a diagram of a Petri net,
places are depicted as circles, bars as straight lines, and arcs as arrows
between the places and bars.

A *marking* of a Petri net is an assignment of numbers of *tokens* to the places
in the net. Intuitively, a marking is a "state" of the net. A marking is depicted by appropriate numbers of black dots in the places of the net.

A bar is said to be *enabled* by a particular marking if every input place to
the bar has at least as many tokens as there are arcs from the place to the
bar. An enabled bar may *fire*, removing one token from each input place for
every arc from the place to the bar and adding a token to each output place

for every arc from the bar to the place. A *firing sequence* is an alternating sequence of markings and bars, beginning with an initial marking. Each marking must enable the bar that follows it in the sequence, and firing that bar must result in the next marking in the sequence. The trace structure consists of all the traces that correspond to the labels on firing sequences of the Petri net.

Our Petri nets have two additional properties. First, each bar is labeled with a wire name. The firing of a bar represents a signal transition on the wire which labels it. Thus, the sequence of wire names which correspond to the bars in a firing sequence represents a possible history of wire transitions. Second, our Petri nets are *bounded*, meaning that only a finite set of markings are reachable from the initial marking. This restriction ensures that the specification is finite-state.

## 7.3. TREE ARBITER

Our first example is an arbiter that has a history of use as an example in discussions of verification of asynchronous circuits. The design was first presented by Seitz in 1980 [14]. Bochmann specified and verified it using temporal logic, without computer assistance, discovered a problem, and proposed a modified design [4]. Using automatic verification, Dill and Clarke discovered a bug in Bochmann's design, and also verified a corrected (and simpler) design by Seitz [5].

The approach used by Dill and Clarke was *temporal logic model checking*. It used a *model checker*, which compares a state graph describing the behavior of the implementation with a temporal logic specification. The state graph was extracted from a structural description of the circuit using a special-purpose pre-processor. Although the model-checker is very fast, the state graph can grow exponentially with the size of the circuit, so the challenge in verifying this circuit was to fit the state graph in available memory. In fact, this was such a serious problem that they were never able to construct the full state graph — instead, they constructed the smaller state graph that resulted when only one user made requests. This was sufficient to detect a problem.

There were two major problems with this approach. First, the complete state graph had to be built before verification could begin. This is often unnecessary because a problem may be evident after relatively few states have been constructed. Moreover, state graphs of incorrect circuits are often much larger than those of correct circuits because of random events occuring after an error. This effect was dramatic in the tree arbiter example: the complete state graph of the incorrect arbiter had more than 11,000 states (at

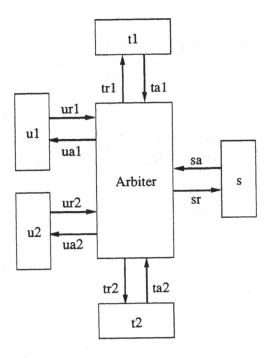

**Figure 7-6.** Block diagram of a tree arbiter.

which point memory was exhausted); that of the corrected arbiter had only 62 states.

Another disadvantage of the model-checking approach is that descriptions of components modelled all possible inputs — including illegal ones (in an asynchronous circuit, many input sequences would cause hazards and should therefore be disallowed). The result is that errors were not detected as quickly as they could have been; generally, a glitch on a gate input would cause hazard on the gate output (which showed up in the state graph as a possible change in the output). The glitch would propagate around inside the circuit until it caused a possible bad output which violated the temporal logic specification.

## 7.3.1. Specification of the Tree Arbiter

An arbiter ensures mutually-exclusive access to a resource. The tree arbiter has five *four-phase interfaces*: two for clients (users), one for a *server* (representing the processing done in the critical region), and two for *transfer modules* which do some setup computation for the server. A block diagram for the arbiter appears in figure 7-7. A four-phase interface consists of two

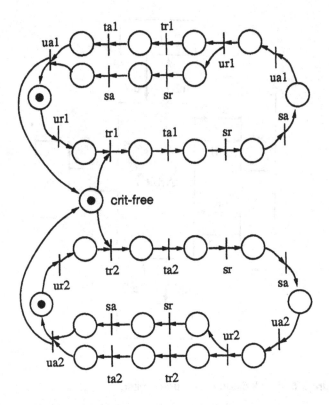

**Figure 7-7.** Petri net specification of a tree arbiter.

wires, *request* and *acknowledge*, and a protocol for using them to negotiate transactions between a client and server. The protocol is: *request* and *acknowledge* are initially at logical 0; (1) client asks for service by raising *request*; (2) server acknowledges by raising *acknowledge*; (3) client lowers *request*; (4) server lowers *acknowledge*; the interface is ready for another request. To ensure mutual-exclusion, the user raises *uri* (for $i = 1,2$) to ask for the resource; the arbiter grants access to the resource by raising *uai*. Then the arbiter is in its *critical region* — at most one user interface is allowed in this phase at any time. The user lowers *uri* to release the resource, and the arbiter acknowledges this by lowering *uai*.

Once the tree arbiter grants the resource (let us say to user 1), it makes a request (on *tr1*) to a transfer module. After this is complete (indicated by *ta1* going high), the arbiter makes a request to the server (the resource in this case). When the server indicates completion by raising *sal*, the arbiter goes through the return-to-zero phases on all relevant interfaces.

A Petri net specification (adapted from Seitz [14]) for this behavior is shown in figure 7-7. There are two main cycles in the net; each represents the processing of a different user's request. If one or more user requests are asserted (*url* or *ur2*), and there is a token in the place *crit-free*, then one such request may be selected and a token cycles through the appropriate sequence of transitions in the net.

## 7.3.2. Bad Implementation of the Arbiter

An incorrect implementation of the arbiter is shown in figure 7-8. This was the design published (and verified correct) by Bochmann after discovering a problem in the original design. The components in the design consist of familiar AND and OR gates, and two elements that are common in asynchronous designs: a *C-element* and a *ME-element*. If both inputs of a C-element are the same, the output eventually goes to the same value. Otherwise, the output does not change. C-elements are often used to wait for several processes to complete. The ME (mutual-exclusion) element is a primitive that resolves meta-stable states: the outputs basically follow the corresponding inputs, except that if both inputs are high, it ensures that only

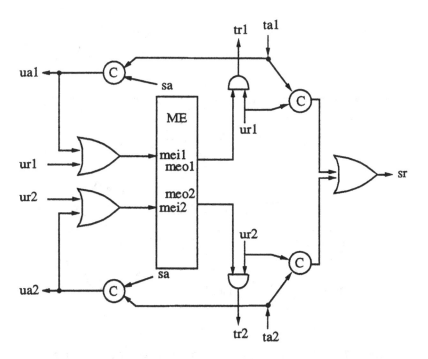

**Figure 7-8.** Implementation of a tree arbiter.

one output (chosen arbitrarily) is high. The Petri nets of figure 7-9 and figure 7-10 describe the behaviors of the C-element and the ME- element, respectively.

To verify this circuit, we constructed trace structures for the gates and elements and the specification by using the Boolean function and Petri net translators. We then wrote an expression (with trace structures for the components as operands and with **rename, hide,** and **compose** as operators) to give a structural description of the circuit. Then we called **conforms-to-p** with this expression and the specification.

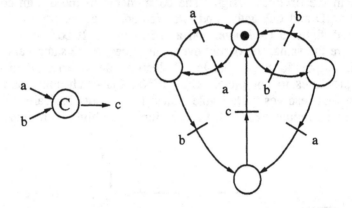

**Figure 7-9.** C-element and its Petri net specification.

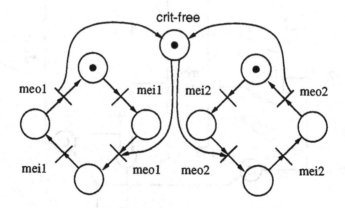

**Figure 7-10.** Petri net specification of a mutual exclusion element.

Using our Petri net translator, the specification was translated into internal form in 0.4 seconds; 60 states were generated. The verifier program was then run; it reported an error after examining 19 states in 0.2 seconds, and printed the following trace:

*ur1 mei1 meo1 tr1 ta1 x1 sr sa ua1 ur1 tr1 ta1 x1 sr sa ua1 ur1.*

The problem was an internal failure in the OR-gate with inputs *ur1* and *ua1*. The user asserted a request *ur1*; the arbiter granted the request, and cycled through the appropriate transitions. It then finally reset the wire *ua1*, allowing a new user request on the same interface. However, in this scenario, the OR-gate's output was never reset. When *ua1* is reset, its inputs are both low while the output remains high. A new request *ur1* may now arrive as an input to the OR-gate, before its output has had a chance to reset; this is a failure according to our treatment of gates.

Notice in this example how the environmental constraint on gates helped to detect the error quickly. Without the environmental constraint, we would have had to model a possible glitch on the output of the gate. This glitch would have propagated through the circuit until it eventually caused an incorrect output, which would have violated the specification of the arbiter. So the environmental constraint on the OR-gate saved the states that would have been needed to model the propagation of the glitch.

### 7.3.3. Correct Arbiter Implementation

A corrected implementation is shown in figure 7-11 [5]. The verifier examines 62 states in 0.1 seconds, and confirms that the implementation is valid. In this case, there were no appreciable savings over the model-checking approach, since all states needed to be examined to prove that there were no errors. However, no savings were necessary — the state graph for the correct circuit was so much smaller than that of the previous implementation that it was easily handled by both programs.

## 7.4. ARBITER WITH REJECT

Our next example is an original design for a speed-independent, four-phase *arbiter with reject*[2]. In a normal arbiter, once an input request is acknowledged by the arbiter, any new request will block (wait) until the other

---

[2] Thanks to Al Davis and Ken Stevens of Hewlett-Packard Laboratories, Palo Alto, for posing the problem of designing this circuit.

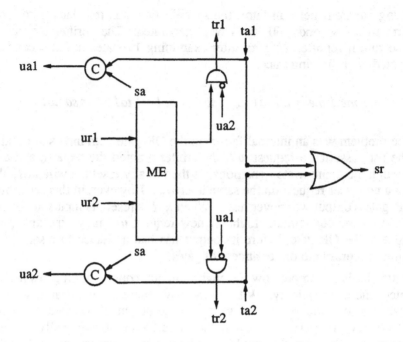

**Figure 7-11.** Corrected implementation of a tree arbiter.

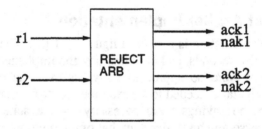

**Figure 7-12.** Block diagram of an arbiter with reject.

user has freed the resource. Sometimes, a requester would prefer to be told that the resource is not available, so that it can do something else instead of waiting. The arbiter with reject has two outputs for each user: *ack*, which grants the resource, and *nak*, which indicates that the resource is in use. A block diagram for an arbiter with reject appears in figure 7-12.

## 7.4.1. Specification

A Petri net specification of the arbiter appears in figure 7-13. The places labeled *crit-free* and *in-crit* are used to model whether a (positive) acknowledgment for any request is currently outstanding. A single token in the place *crit-free* indicates that no acknowledgment is outstanding (the shared resource is free); a token in *in-crit* indicates that a request is currently acknowledged (the resource is being used).

The left and right subnets enforce the normal four-phase handshaking protocols for the first and second interfaces, respectively. Some time after a request has been asserted, it will either be *ack'd* or *nak'd* depending on whether or not the other interface is in its critical region.

## 7.4.2. Implementation

A block diagram for the design is shown in figure 7-14. The output modules control the setting of *ack* and *nak* for each interface. A signal on *rqst-set* is a "request to set *ack*". The output module responds to such a request by setting, say, *ack1* if *ack2* is not already set. Otherwise, it sets *nak1*. Then the request is acknowledged on *valid-out*. A second request is "request to reset *ack* or *nak*." It causes whichever signal is high to go low again.

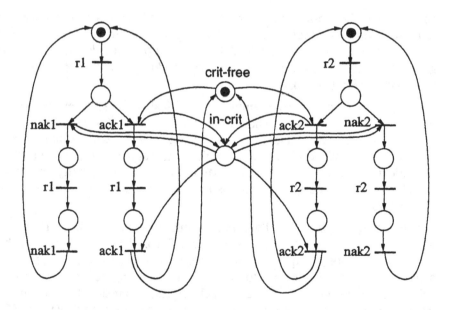

**Figure 7-13.** Petri net specification of an arbiter with reject.

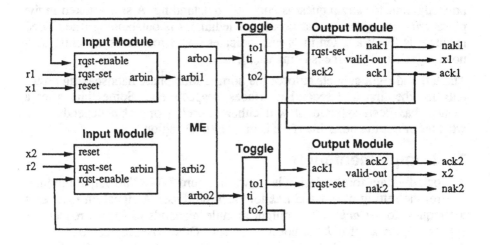

**Figure 7-14.** Block diagram for design of arbiter with reject.

Unfortunately, there is a race condition: what if *ack2* is changing at the same time that the other module is deciding whether to set *ack1* or *nak1*? This could result in a metastable state in the output module. It might also generate a runt pulse on the output of the module, or both *ack1* and *ack2* might be set simultaneously.

We avoid this problem by defining two critical regions: when an output module is being set, and when it is being reset. Mutual exclusion is enforced by the ME-element. We manage to get by with a single ME-element to arbitrate between four possible combinations of input requests by multiplexing these requests (the ME-element is essentially working "double-time"). The multiplexing is implemented using the *input module* and *toggles*.

A Petri net specification for the input module appears in figure 7-15, and a hardware implementation appears in figure 7-16. The three wires on the input interface obey simple transition-signalling. A high value on *rqst-set* is a request to the ME-element to set the corresponding *ack or nak*; a low value is a request to reset. For each such request, the rest of the implementation can use the two other inputs, *rqst-enable* and *reset* to drive the ME-element through its four-phase cycle; the *rqst-enable* is also used to protect the current ME input from the user's next request. (The ME-element in

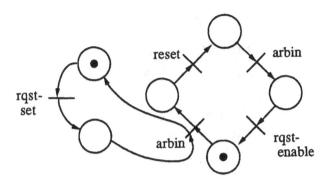

**Figure 7-15.** Petri net for the arbiter input module.

figure 7-14 is the same as the one described in figure 7-10, but with its wires renamed.)

We use simple transition-signalled *toggle elements* on the ME outputs to route the odd-numbered transitions, which are requests to set/reset *ack* or *nak* to the output module, and the even transitions, which acknowledge that the ME has been reset, back to the input module. Each transition on the toggle element input causes a corresponding transition on one of its two outputs. These output transitions alternate. A Petri net for the toggle is shown in figure 7-17. The toggle input is labeled *ti*; the two outputs are *to1* and *to2*.

A Petri net specification for this output module appears in figure 7-18, and a hardware implementation appears in figure 7-19. This module implementation uses a four-phase handshaking. (The circuit has an interesting mix of two- and four-phase components.) The module accepts a request, *rqst-set*, to set either the corresponding *ack* or *nak*. Once the request is made the module then samples the *ack* of the *other* interface, and produces the appropriate output (*ack* or *nak*). It then generates a completion signal,

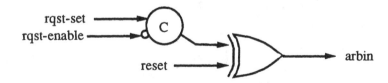

**Figure 7-16.** Input module implementation.

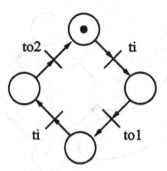

**Figure 7-17.** Petri net specification of the toggle element.

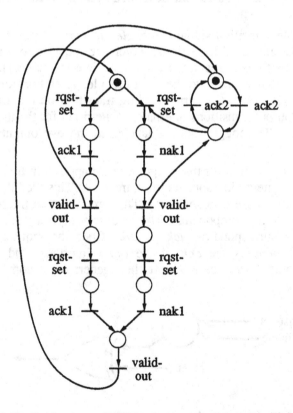

**Figure 7-18.** Petri net specification of output module.

*valid-out*, at which time this other *ack* value may change arbitrarily. A similar resetting cycle occurs as well.

### 7.4.3. Flat Verification

We verify this implementation using two different approaches. First, the
implementation as a whole can be verified directly against the specification.
Since this flattened implementation has many internal interfaces, the verifier
must examine a relatively large number of states.

The complete verification process has two parts. First, the Petri net
specification for the arbiter is converted into an internal representation. This
conversion produces 28 states in 0.2 seconds.

Next, an expression describing the flattened circuit implementation is
verified against this specification (the component trace structures are
predefined in a library, since they are all standard). The verifier examines
2040 states in 7.6 seconds; it confirms that the implementation of the arbiter
is a valid speed-independent implementation.

### 7.4.4. Hierarchical Verification

The verification can also be carried out hierarchically. This has two advantages: the individual problems are smaller, and the results for repeated
modules can be re-used. It has been proved that if a circuit has been shown
to be correct by hierarchical verification, "flat" verification of the same circuit will yield the same result [2].

The hierarchical verification of the arbiter requires two steps. First, a
description corresponding to the block diagram of figure 7-14, which treats

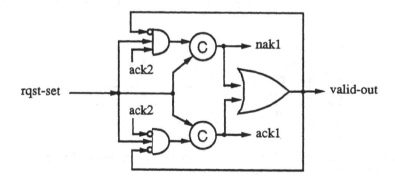

**Figure 7-19.** Output module implementation.

the input and output modules as primitive components, is verified against the arbiter specification. The verifier examines 675 states in 1.0 seconds and finds no problems.

Next, the implementations of both the input and output modules (figure 7-16 and figure 7-19) must themselves be verified against their specifications (figure 7-15 and figure 7-18).

Conversion of the input-module specification produces 8 states and takes 0.1 seconds. The output-module conversion produces 16 states in 0.1 seconds. Finally, each module implementation is verified against its formal specification. The input-module implementation is verified in 0.1 seconds; 18 states are examined. The output-module implementation is likewise verified against its specification in 0.1 seconds; 26 states are examined.

## 7.5. PERFORMANCE SUMMARY

We have verified three examples. There were two major computations involved: (1) converting Petri net specifications into trace structures, and (2) comparing the trace structures for the implementation and specification using **conforms-to-p**. Table 7-1 shows the number of states and time for each conversion of a Petri net specification to a trace structure. Table 7-2 summarizes the number of states created and time for each application of **conforms-to-p**. (not including Petri-net conversions).

We chose this arbiter as an example because it seems unlikely that an automatic synthesis tool would generate the implementation we give here. When custom design is essential (for efficiency or other reasons), it is extremely helpful to have an automatic verification tool to detect any design errors. Verification not only assures the correctness of the final design, it expedites the implementation by quickly finding problems in preliminary

| Examples | States | Time (secs) |
|---|---|---|
| Tree arbiter | 60 | 0.4 |
| Rejecting arbiter | 28 | 0.2 |
| Input module | 8 | 0.1 |
| Output module | 16 | 0.1 |

**Table 7-1.** The number of states and time for each conversion of a Petri net specification to a trace structure.

| Example | States | Time (secs) |
|---|---|---|
| Wrong tree arbiter | 19 | 0.2 |
| Correct tree arbiter | 62 | 0.1 |
| Flat rejecting arbiter | 2040 | 7.6 |
| Hierarchical rejecting arbiter (total) | 719 | 1.2 |
| Top-level design | 675 | 1.0 |
| Input module | 18 | 0.1 |
| Output module | 26 | 0.1 |

**Table 7-2.** The number of states created and time for each application of the conformation procedure.

versions. It also promotes better designs by reducing the risk of introducing errors while improving a circuit. These considerations are especially important in self-timed circuit design because it is often more subtle than synchronous design.

There are limits to the power of trace theory, however. Trace theory cannot express liveness properties, so deadlocks and livelocks may go undetected even in circuits that have been verified. Also, it is often not very suitable for modeling or specifying computations on data. Finally, it is important to be able to deal with circuits that are not speed-independent (whose correct behavior depends on timing) as most practical circuits do rely on timing assumptions for a more efficient implementation.

## REFERENCES

1.    D. L. Dill, "Trace Theory for Automatic Hierarchical Verification of Speed-Independent Circuits," *Proc. of Fifth MIT Conf. on Advanced Research in VLSI*, (1989).

2.    D. L. Dill, *Trace Theory for Automatic Hierarchical Verification of Speed-Independent Circuits*, MIT Press (1989).

3.    D. L. Dill and R. F. Sproull, "Automatic Verification of Speed-Independent Circuits with Petri Net Specifications," *Proc. of IEEE ICCD-89*, (October 1989).

4.    G. V. Bochmann, "Hardware Specification with Temporal Logic: An Example," *IEEE Trans. on Computers* C-31(3) pp. 223-231 (March, 1982).

5.    D.L. Dill and E.M. Clarke, "Automatic Verification of Asynchronous Circuits Using Temporal Logic," *IEE Proceedings* 133(5) pp. 276-282

(September 1986).

6. C. A. R. Hoare, "A Model for Communicating Sequential Processes," *Programming Research Group, Oxford University Computing Laboratory, No. PRG-22*, (1981).

7. M. Rem, J. L. A. van de Snepscheut, and J. T. Udding, "Trace Theory and the Definition of Hierarchical Components," *Third CalTech Conference on Very Large Scale Integration*, pp. 225-239 (April, 1983).

8. J. E. Hopcroft and J. D. Ullman, *Introduction to Automata Theory, Languages, and Computation*, Addison-Wesley Publishing Company (1979).

9. J. L. Peterson, *Petri Net Theory and the Modeling of Systems*, Prentice-Hall (1981).

10. T.-A. Chu, "On the Models for Designing VLSI Asynchronous Digital Systems," *INTEGRATION The VLSI journal*, (4) pp. 99-113 (1986).

11. T. H.-Y. Meng, R. W. Brodersen, and D. G. Messerschmitt, "Automatic Synthesis of Asynchronous Circuits from High Level Specifications," *IEEE Trans. on CAD* CAD-8(11) pp. 1185-1205 (November 1989).

12. C. E. Molnar, T.-P. Fang, and F. U. Rosenberger, "Synthesis of Delay-Insensitive Modules," *1985 Chapel Hill Conf. on Very Large Scale Integration*, pp. 67-86 (1985).

13. S. S. Patil, "An Asynchronous Logic Array," *Massachusetts Institute of Technology, Project MAC, Technical Memorandom 62*, (1975).

14. C. L. Seitz, "Ideas about Arbiters," *Lambda* **First Quarter** pp. 10-14 (1980).

# INDEX